Lecture Notes in Computer Science 14517

Founding Editors

Gerhard Goos
Juris Hartmanis

Editorial Board Members

Elisa Bertino, *Purdue University, West Lafayette, IN, USA*
Wen Gao, *Peking University, Beijing, China*
Bernhard Steffen, *TU Dortmund University, Dortmund, Germany*
Moti Yung, *Columbia University, New York, NY, USA*

The series Lecture Notes in Computer Science (LNCS), including its subseries Lecture Notes in Artificial Intelligence (LNAI) and Lecture Notes in Bioinformatics (LNBI), has established itself as a medium for the publication of new developments in computer science and information technology research, teaching, and education.

LNCS enjoys close cooperation with the computer science R & D community, the series counts many renowned academics among its volume editors and paper authors, and collaborates with prestigious societies. Its mission is to serve this international community by providing an invaluable service, mainly focused on the publication of conference and workshop proceedings and postproceedings. LNCS commenced publication in 1973.

Michael Harrison · Célia Martinie ·
Nicholas Micallef · Philippe Palanque ·
Albrecht Schmidt · Marco Winckler ·
Enes Yigitbas · Luciana Zaina
Editors

Engineering Interactive Computer Systems

EICS 2023 International Workshops and Doctoral Consortium

Swansea, UK, June 26–27, 2023
Selected Papers

Editors
Michael Harrison
Newcastle University
Newcastle upon Tyne, UK

Nicholas Micallef
Swansea University
Swansea, UK

Albrecht Schmidt
Ludwig-Maximilians-Universität München
München, Germany

Enes Yigitbas
University of Paderborn
Paderborn, Germany

Célia Martinie
Paul Sabatier University
Toulouse, France

Philippe Palanque
IRIT - Interactive Critical Sys Group
Paul Sabatier University
Toulouse, France

Marco Winckler
Université Côte d'Azur
Sophia Antipolis Cedex, France

Luciana Zaina
Federal University of São Carlos
Sorocaba, Brazil

ISSN 0302-9743   ISSN 1611-3349 (electronic)
Lecture Notes in Computer Science
ISBN 978-3-031-59234-8   ISBN 978-3-031-59235-5 (eBook)
https://doi.org/10.1007/978-3-031-59235-5

© The Editor(s) (if applicable) and The Author(s), under exclusive license to Springer Nature Switzerland AG 2024

This work is subject to copyright. All rights are solely and exclusively licensed by the Publisher, whether the whole or part of the material is concerned, specifically the rights of translation, reprinting, reuse of illustrations, recitation, broadcasting, reproduction on microfilms or in any other physical way, and transmission or information storage and retrieval, electronic adaptation, computer software, or by similar or dissimilar methodology now known or hereafter developed.
The use of general descriptive names, registered names, trademarks, service marks, etc. in this publication does not imply, even in the absence of a specific statement, that such names are exempt from the relevant protective laws and regulations and therefore free for general use.
The publisher, the authors and the editors are safe to assume that the advice and information in this book are believed to be true and accurate at the date of publication. Neither the publisher nor the authors or the editors give a warranty, expressed or implied, with respect to the material contained herein or for any errors or omissions that may have been made. The publisher remains neutral with regard to jurisdictional claims in published maps and institutional affiliations.

This Springer imprint is published by the registered company Springer Nature Switzerland AG
The registered company address is: Gewerbestrasse 11, 6330 Cham, Switzerland

If disposing of this product, please recycle the paper.

Foreword

This book presents a series of revised papers selected from the Doctoral Consortium (DC) and the Workshops organized in conjunction with the 15th ACM SIGCHI Symposium on Engineering Interactive Computing Systems (EICS 2023), which was held in Swansea, in Wales, UK, in June 2023. Even though EICS itself has a long track record, this is the first time that papers submitted to DC and workshops at EICS are collected in a single volume.

The symposium on Engineering Interactive Computing Systems (EICS) is the primary venue for research contributions at the intersection of Human-Computer Interaction (HCI) and Software Engineering. EICS is organized by the Special Group on Human-Computer Interaction of the Association for Computing Machinery (ACM SIGCHI) and the Working Group on User Interface Engineering (WG 2.7/13.4) of the International Federation for Information Processing (IFIP). EICS 2023 was the fifteenth edition of a series of conferences devoted to engineering interactive computing systems and their user interfaces. The conference is best known for rigorously contributing and disseminating research results that occupy the middle ground between user interface design, software engineering, and computational interaction. EICS focuses on models, languages, notations, methods, techniques, and tools that support designing, developing, validating, and verifying interactive systems. Whilst such a focus might concern a large spectrum of application domains, EICS also encourages and promotes by the means of specialized workshops research on emerging topics related to engineering interactive systems.

Workshop proposals submitted to EICS are juried. Selection criteria include the relevance of the workshop topics and goals to EICS and the scientific background and experience of the workshop's organizers in organizing events, advertising, and a dissemination plan. For EICS 2023, Philippe Palanque, Albrecht Schmidt, Enes Yigitbas, and Nicholas Micallef were responsible for the selection process, which resulted in two full-day workshops held in parallel on June 27–30, 2023:

- **Engineering Interactive Computing Systems for People with Disabilities (DISAB 2023).** The aims of this workshop were to explore the development and use of interactive systems engineering for users with disabilities. This workshop welcomed contributions that would help to overcome permanent disabilities (such as visual, hearing, mobility impairments...), evolutive (in the case of degenerative diseases such as Alzheimer's and Parkinson's) or temporary (situationally impaired people).
- **Engineering Interactive Systems Embedding AI Technologies (EIS-embedding-AI).** The main goal of this workshop was to offer a platform for scientists who are interested in the design, development, evaluation, and use of interactive systems involving AI technologies. Two objectives were settled: to identify and gather information about knowledge and practice in the workshop's domain; and elicit the main gaps in AI technologies which hinder their exploitation in the design and development of interactive systems, especially if a user-centered design process is followed.

Each workshop had its own program committee who oversaw the selection process of contributions to be presented in Swansea. Following the discussions during the workshops, authors were then invited to revise their work and submitted the contribution for a second round of reviews. At the end, twelve papers (six from each workshop) were selected for inclusion in this book featuring the post-proceedings of the EICS 2023 workshops, which also includes five contributions submitted to the Doctoral Consortium.

Contributions submitted to the EICS 2023 Doctoral Consortium (DC) were juried by a committee of senior researchers under the leadership of Michael Harrison, Célia Martinie, and Luciana Zaina. The DC was an opportunity for doctoral students to present their on-going work and to discuss their results with leading experts in the field, as well as with peers. Students attending the EICS 2023 DC were at intermediate and advanced stages of their research. The students that presented substantial material at the DC were then invited to prepare a full paper that, after a careful reviewing process, was considered for publication in these post-proceedings.

The doctoral consortium papers provide fascinating insights into current, new research introducing new themes to the engineering of interactive computer systems. The work is inevitably at different stages of maturity, showing the developing field and the progress of the community. Four of the five papers are about providing software support to a process, aiming to bring the approaches available to support the development of interactive systems to a wider community. Two of the papers (Artizzu and Calo) are about supporting the development of novel systems: extended reality and making sense of sketches to produce code. Macedo's work concerns a stage further back in the process, enabling software engineers to make sense of interaction data, aiding understanding of the role of the data in improving design. Moreira's work is designed to help a software engineer make sense of the often difficult to interpret results of a proof when model checking the formal description of an interactive system. Finally, Gomez-Monroy provides a fascinating example of the use of smartphones to help children make sense of and perform "boxing moves" effectively.

Altogether, the work is promising and encouraging, and we look forward to seeing further developments from the authors.

Michael Harrison
Célia Martinie
Marco Winckler

Organization

General Chair

Alan Dix   Swansea University, UK

Technical Program Chairs

Marco Winckler   Université Côte d'Azur, France
Matt Jones   Swansea University, UK

Full Paper and Technical Notes Chairs

Carmen Santoro   ISTI-CNR, Pisa, Italy
Kris Luyten   Hasselt University, Belgium

Late-Breaking Results Chairs

Anke Dittmar   University of Rostock, Germany
Simon Robinson   Swansea University, UK

Workshops Chairs

Albrecht Schmidt   LMU Munich, Germany
Enes Yigitbas   University of Paderborn, Germany
Nicholas Micallef   Swansea University, UK
Philippe Palanque   Université Toulouse 3, France

Doctoral Consortium Chairs

Célia Martinie   Université Toulouse III - Paul Sabatier, France
Luciana Zaina   Federal University of São Carlos - UFSCar, Brazil
Michael Harrison   Newcastle University, UK

Tutorials Chairs

Jean Vanderdonckt — Université catholique de Louvain, Belgium
Pranjal Jain — Swansea University, UK
Simone Barbosa — PUC-RIO, Brazil

Demo and Posters Chairs

Anna Carter — Swansea University, UK
Marc Hesenius — University of Duisburg-Essen, Germany

Publicity Chair

Benjamin Weyers — University of Trier, Germany

Technical and Accessibility Chair

Jennifer Pearson — Swansea University, UK

SIGCHI/ACM Liaison

Judy Bowen — Waikato University, New Zealand

SIGCHI CARES Representative

Jennifer Pearson — Swansea University, UK

Web Site

Gavin Bailey — Swansea University, UK

Steering Committee Chair

José Creissac Campos — Universidade do Minho, Portugal

Steering Committee Members

Aaron Quigley	University of New South Wales, Australia
Alan Dix	Swansea University, UK
Gaëlle Calvary	Université Grenoble Alpes, France
Jeff Nichols	Apple, USA
José C. Campos	Universidade do Minho, Portugal (committee chair)
Judy Bowen	University of Waikato, New Zealand
Kris Luyten	Hasselt University, Belgium (previous committee chair)
Marco Winckler	Université Côte d'Azur, France
Panos Markopoulos	Eindhoven University of Technology, Netherlands
Philippe Palanque	Univ. Toulouse 3, France (previous committee chair)

Associated Chairs

Anke Dittmar	University of Rostock, Germany
Arnaud Blouin	University of Rennes, INSA Rennes, France
Carmen Santoro	CNR-ISTI, Italy
Davy Vanacken	Hasselt University - tUL - Flanders Make, Belgium
Jeffrey Nichols	Apple Inc, USA
Jürgen Ziegler	University of Duisburg-Essen, Germany
Fabio Paternò	CNR-ISTI, Italy
Gustavo Rovelo Ruiz	Hasselt University, Belgium
José Creissac Campos	Universidade do Minho, Portugal
Judy Bowen	University of Waikato, New Zealand
Kris Luyten	Hasselt University - tUL - Flanders Make, Belgium
Lucio Davide Spano	University of Cagliari, Italy
Mathias Funk	Eindhoven University of Technology, Netherlands
Michael Harrison	Newcastle University, UK
Peter Forbrig	University of Rostock, Germany
Philippe Palanque	Université Toulouse 3, France
Radu-Daniel Vatavu	University of Suceava, Romania
Sophie Dupuy-Chessa	Université Grenoble Alpes, France

Additional Reviewers

Aditya Shekhar Nittala	University of Calgary, Canada
Adrien C. Caillet	Laboratoire d'Informatique de Grenoble, France
Alan Dix	Swansea University, UK
Albrecht Schmidt	LMU Munich, Germany
Alexandre Demeure	Université Grenoble Alpes, France
Aliaksei Miniukovich	University of Innsbruck, Austria
André Zenner	German Research Center for AI (DFKI), Germany
Andrea Mattioli	CNR-ISTI, Italy
Antti Oulasvirta	Aalto University, Finland
Audrey Sanctorum	Vrije Universiteit Brussel, Belgium
Carlos Silva	Centro de Computação Gráfica, Portugal
Célia Martinie	IRIT, Université Toulouse III - Paul Sabatier, France
Chen Guo	James Madison University, USA
Cristian Pamparau	Ștefan cel Mare University of Suceava, Romania
Christian Maertin	Augsburg University of Applied Sciences, Germany
Daniel Prun	École Nationale de l'Aviation Civile, France
Danny Leen	Hasselt University, Belgium
David Navarre	University of Toulouse, France
Donald Degraen	German Research Center for AI (DFKI), Germany
Eldon Schoop	University of California, Berkeley, USA
Eleonora Zedda	HIIS Laboratory, CNR-ISTI, Italy
Elise Lavoué	Université Jean Moulin Lyon 3, France
Emilia Barakova	Eindhoven University of Technology, Netherlands
Emilyann Nault	Heriot-Watt University, UK
Eva Geurts	UHasselt - tUL - Flanders Make, Belgium
Florian Echtler	Aalborg University, Denmark
Florian Heller	Hasselt University - tUL - Flanders Make, Belgium
Forrest Huang	Apple, USA
Francisco Maria Calisto	Universidade de Lisboa, Portugal
Haijun Xia	University of California San Diego, USA
Hsin-Ruey Tsai	National Chengchi University, Taiwan
Jan Van den Bergh	UHasselt - tUL - Flanders Make, Belgium
Jason Wu	Carnegie Mellon University, USA
Jean-Sebastien Sottet	Luxembourg Institute of Science and Technology, Luxembourg
Jemma Konig	University of Waikato, New Zealand
Jessica Turner	University of Waikato, New Zealand

Jo Vermeulen	Autodesk Research, Canada
José Rouillard	Université de Lille, France
Julius von Willich	TU Darmstadt, Germany
Kathia Oliveira	Université Polytechnique Hauts-de-France, France
Klen Čopič Pucihar	University of Primorska, Slovenia
Luigi De Russis	Politecnico di Torino, Italy
Manas Satish Bedmutha	University of California San Diego, USA
Mannu Lambrichts	Hasselt University, Belgium
Marc Hesenius	University of Duisburg-Essen, Germany
Marco Manca	CNR-ISTI, Italy
Maristella Matera	Politecnico di Milano, Italy
Mengyu Chen	University of California Santa Barbara, USA
Min Zhang	Open University, UK
Paolo Masci	NIA/NASA LaRC, USA
Pawel Wozniak	Chalmers University of Technology, Sweden
Pedro A. Santos	Polytechnic Institute of Setúbal, Portugal
Peter Bago	Corvinus University of Budapest, Hungary
Sandip Chakraborty	Indian Institute of Technology Kharagpur, India
Shamus Smith	Griffith University, Australia
Simone Barbosa	PUC-RIO, Brazil
Sophie Lepreux	Université Polytechnique - Hauts de France, France
Stephane Conversy	ENAC, France
Sybille Caffiau	Université Grenoble Alpes, France
Veronika Krauss	University of Siegen, Germany
Vivian Genaro Motti	George Mason University, USA
Wei Bai	Google LLC, USA
Yann Savoye	Liverpool John Moores University, UK
Yevgen Borodin	Stony Brook University, USA
Yin Li	Cornell University, USA
Yue Jiang	Aalto University, Finland
Yosra Rekik	Université Polytechnique Hauts-de-France, France
Zhu Wang	Future Reality Lab, USA

Contents

Engineering Interactive Computing Systems for People with Disabilities (DISAB 2023 Workshop)

A First Literature Study on Predictive Quality in Use Evaluation for Smart Environments 3
  Maria Paula Corrêa Angeloni, Káthia Marçal de Oliveira, and Emmanuelle Grislin-Le Strugeon

A First Step Towards an Ecosystem Meta-model for Human-Centered Design in Case of Disabled Users 11
  Christophe Kolski, Nadine Vigouroux, Yohan Guerrier, Frédéric Vella, and Marine Guffroy

Evaluation of a Social Robot System for Performance-Oriented Stroke Therapy 20
  Alexandru Umlauft and Peter Forbrig

MUMR-MIODMIT: A Generic Architecture Extending Standard Interactive Systems Architecture to Address Engineering Issues for Rehabilitation 28
  Axel Carayon, Célia Martinie, and Philippe Palanque

Serious Game for Company Governance: Supporting Integration, Prevention of Professional Disintegration and Job Retention of People with Disabilities 41
  Yosra Mourali, Benoit Barathon, Maxime Bourgois Colin, Sondes Chaabane, Raja Fassi, Alice Ferrai, Yohan Guerrier, Dorothée Guilain, Christophe Kolski, Yoann Lebrun, Sophie Lepreux, Philippe Pudlo, and Jason Sauvé

Two Concepts of Domain-Specific Languages for Therapists to Control a Humanoid Robot 50
  Peter Forbrig, Alexandru Umlauft, Mathias Kühn, and Anke Dittmar

Engineering Interactive Systems Embedding AI Technologies (EIS-embedding-AI Workshop)

An Approach to Leverage Artificial Intelligence for Car-Parking Related Mobile Applications 63
  Alba Bisante, Venkata Srikanth Varma Datla, Gabriella Trasciatti, Stefano Zeppieri, and Emanuele Panizzi

Engineering AI-Similar Designs: Should I Engineer My Interactive System with AI Technologies? 72
    David Navarre, Philippe Palanque, and Célia Martinie

Explaining Through the Right Reasoning Style: Lessons Learnt 90
    Lucio Davide Spano and Federico Maria Cau

Exploring AI-Enhanced Shared Control for an Assistive Robotic Arm 102
    Max Pascher, Kirill Kronhardt, Jan Freienstein, and Jens Gerken

Hidden Figures: Architectural Challenges to Expose Parameters Lost in Code 116
    Alan Dix

Not What I was Trained for – Out-of-Distribution-Tests for Interactive AIs 127
    Benedikt Severin, Ole Werger, and Marc Hesenius

Doctoral Consortium EICS 2023

End User Development for Extended Reality 151
    Valentino Artizzu

Exertion Trainer: Smartphone Exergame Design to Support Children's Kinesthetic Learning Through Playful Feedback 166
    Carla Gómez-Monroy and Alejandro C. Ramírez-Reivich

Explaining Temporal Logic Model Checking Counterexamples Through the Use of Structured Natural Language 179
    Ezequiel José Veloso Ferreira Moreira and José Creissac Campos

Merging Creativity with Computation in Sketch-to-Code Transitions 198
    Tommaso Calò

UX Data Visualization: Supporting Software Professionals in Exploring Users' Interaction Data 207
    Maylon Macedo and Luciana Zaina

Author Index 223

Engineering Interactive Computing Systems for People with Disabilities (DISAB 2023 Workshop)

A First Literature Study on Predictive Quality in Use Evaluation for Smart Environments

Maria Paula Corrêa Angeloni[1(✉)], Káthia Marçal de Oliveira[1], and Emmanuelle Grislin-Le Strugeon[1,2]

[1] UPHF, CNRS, UMR 8201 - LAMIH, Famars, France
mpangeloni@gmail.com
[2] INSA Hauts-de-France, Famars, France

Abstract. Smart environments promote an habitat where technological devices may facilitate the daily tasks of whoever lives in them, which may provide a good living alternative for the elderly or for people with some kind of motor impairment, therefore, it is essential to measure the quality of the software applications found in such places. In this paper, we conduct a review to analyse studies about the prediction of software quality and initiate the design of our approach to predict the Quality in Use of applications in smart environments without having to involve end users. Instead, we intend to implement a multi-agent system to simulate human behaviour, mine the data from the simulation and apply measures from a ISO/IEC published standard to predict the software quality.

Keywords: Quality in Use · Multi-agent systems · Data mining

1 Introduction

Currently, it is common to talk about smart environments (whether they are smart homes, smart cities, or so on) that are available to users for the purposes of assisting their daily activities. Such help might be even better employed for those that have some difficulties while performing day-to-day tasks and may need adapted spaces to accommodate their needs, such as elderly people, who may face more challenges when doing certain activities because of aging, or people who have some sort of motor disability.

The world population has been growing older thanks to advancements in different fields, as medicine and technology, which increases the number of elderly citizens (considered as such as of the age of 65) [21]. One alternative to improve the life quality and safety of the senior population is to create "ageing-friendly living environments with the application of smart home technologies" [15], and mitigate the absence of caregiving with technological features while promoting a possible solution from not only a practical, but also economic perspective [7]. To ensure the quality of the software in a smart environment, it is necessary to test it and evaluate how it responds to user engagement, measuring how well it works and if it is good enough to fulfill the users' needs regarding such application. Guaranteeing the reliability of tools is even more important in cases where the user is elderly or a person with disabilities, considering a software flaw might put them in danger, specially when it comes to a person with reduced mobility.

According to the ISO/IEC 25010 standard [11], the "quality of a system is the degree to which the system satisfies the stated and implied needs of its various stakeholders, and thus provides value". These needs are represented in the standard in two models that categorize product quality into characteristics that are divided into subcharacteristics. The models are: (a) the system/software product quality model that is focused on internal and external properties (such as maintainability or reliability), and (b) a Quality in Use model to measure the users' perceived quality when using the system in specific contexts of use. We will focus on the second one, considering our goal is to assure the quality of software applied specifically in smart environments and how users with some kind of motor impairment can handle it, as we consider that such habitat creates an appropriate alternative to solve the issue of ageing safely at home and also receiving the necessary care.

Besides being considered economically viable, getting older while using smart home technologies is believed to promote advantages for the user's well-being, even though some elderly people are oblivious to their current health status and might have the wrong perceptions about the technological devices that are present in a smart environment [15], not realizing the benefits such place may provide from the beginning. Smart home technologies allow senior citizens and people with disabilities to receive medical support as needed at their convenience, making it possible for them to live in their own conditions and in an independent manner [7]. For this reason, we would like to create a method for predicting software quality instead of placing actual people to try the application out, as putting them in a smart apartment for testing purposes might be difficult or even dangerous if the software has not been tested at all yet. In this paper, we perform a review to analyse if the prediction of software quality has been studied and how it has been applied, besides perceiving how our research could be original and helpful to society. Moreover, we investigate how we can design an approach employing a multi-agent system to simulate human behaviour in a smart apartment, since we believe that due to their characteristics, such places can be naturally modelled and simulated by agent-based techniques that exhibit similar characteristics of distribution, interaction and autonomy [4]. That way, with software agents, we can simulate human actions and their interactions with applications in a smart environment. Furthermore, given that we can generate enough data from a simulation (for example, considering the same user using the application at different times, or several users of different profiles using the application), mining techniques [10] could be applied. This paper presents our first steps towards the definition of this approach.

This paper is organized as follows: Sect. 2 presents a background for our work, providing information about the standard in which we intend to apply our evaluation and with which AI techniques. Section 3 explains the research we performed and the steps that were taken for such. In Sect. 4, we discuss about the results from the performed review, and finally, Sect. 5 presents our final remarks and ongoing work.

2 Background

According to the ISO/IEC 25010 standard [11], Quality in Use (QinU) is the range in which a software can be used according to the user's needs and can reach certain goals related to effectiveness, efficiency, satisfaction, freedom from risk and context coverage. The same standard states that software quality characteristics are quantified by measures that are calculated by considering data collected from the user's interaction with the application. In a smart environment, this data comes from both implicit user interaction (e.g., sensors) and explicit user interaction (e.g., clicking a button in a device). This perspective considers the adaptability to the context as one of its characteristics.

The QinU model includes five characteristics: (i) effectiveness, as the accuracy and completeness with which users achieve specified goals; (ii) efficiency, defined by the resources used in relation to the accuracy and completeness with which users achieve goals; (iii) freedom from risk, which represents how much a product or system mitigates the potential risk to economic status, human life, health, or the environment; (iv) satisfaction, for the degree to which user needs are satisfied when a product or system is used in a specified context of use and by means of the sub-characteristics of comfort, pleasure, trust, and usefulness; (v) context coverage, which is the degree to which a product or system can be used with effectiveness, efficiency, freedom from risk, and satisfaction in both specified contexts of use (context completeness) and contexts beyond those initially explicitly identified (flexibility).

QinU evaluation is even more challenging in smart environments, where the user's context and task are related to an implicit human-computer interaction and not the usual explicit interaction where the user should explicitly make an input in the software system.

Our approach to predict the QinU in the context of smart environments is based on agent mining, which is the combination of agent methods [23] and data mining techniques, such as machine learning [10]. Agent-based modeling and simulation (ABMS) [4] can be used to reproduce interactions between entities that compose complex systems. ABMS has been used, for example, to evaluate building performance with occupant behavior representations [14], to simulate the evacuation in smart buildings [22], or to optimize shared energy resources based on a simulation of human activities [2]. However, we would like to analyse if the particular case of predicting the quality of interactions between humans and smart environments has been studied yet. This would involve modelling and simulating the behaviour of the elements composing the smart environment according to the actions of humans sharing this environment, then using the simulation data to predict the quality from these interactions from a human's point of view. Among the previous works that make use of agent approaches to model and simulate some human performances, approaches based on intentions [1] and needs [9,12] seem appropriate regarding smart environments context.

3 Rapid Review

In order to design our proposal for QinU prediction in smart environments, we decided to carry out a Rapid Review [19] to study the state of the art and analyse how other works have been predicting software quality.

Rapid Review (RR) [19] is a type of knowledge synthesis in which Systematic Literature Review (SLR) processes are accelerated and methods are streamlined to complete the review more quickly than in typical SLRs. For example, RRs may limit search sources or use a single person to select primary studies. The goal is to produce results in a timely manner, with lower costs and, therefore, more connected to practice than a traditional SLR, in which the required time frame may not meet the needs of some decision makers [5,20]. A RR is described as being "within the family" of a SLR, because its methodology has been established as a detailed and transparent scientific method that is reproducible by other people [19]. For these reasons, we decided to perform a RR instead of a classic SLR [13]. In this section, we present the definition of the research protocol (Sect. 3.1) and its execution (Sect. 3.2).

3.1 Planning

Research Questions: From the main research question presented in the introduction, and considering the context of smart/intelligent/pervasive environments, we derived other more specific research questions (according to the AI techniques we wish to employ in our study) as follows:

- RQ1: Is there predictive QinU studies for smart/intelligent/pervasive environments?
- RQ2: Are agents/multi-agent systems used in predictive QinU evaluations?
- RQ3: Is data mining used in predictive QinU evaluations?
- RQ4: Which quality characteristics were evaluated? With what measures?

Search String: The search string was defined using PICOC strategy [17]: *P*opulation, for all studies related to QinU, as well as all its subcharacteristics; *I*ntervention, for finding predictive assessments, which is the focus of this study, considering synonyms for "prediction"; *C*omparison, which was not considered since we did not know of any other reviews with the same goal; *O*utcome, to set that either multi-agent systems or data mining need to be involved in the results; and finally, *C*ontext, to include smart environments, pervasive systems and its synonyms. The subsets of string are then connected by an AND connector. Therefore, the search string was established as follows:

- **Population:** ("usability" OR "quality in use" OR "effectiveness" OR "efficiency" OR "satisfaction" OR "freedom from risk" OR "context*")

- **Intervention (main goal):** ("predictive" OR "prediction" OR "prognostic" OR "prognosticative" OR "anticipating")
- **Outcome**: (("agent" OR "multiagent") OR ("mining" OR "knowledge discovery" OR "KDD"))
- **Context**: ("smart*" OR "intelligent environment" OR "pervasive" OR "ubiquitous" OR "AAL" OR "ambient intelligence" OR "assisted living" OR "systems of systems" OR "internet of things" OR "Cyber-Physical Systems" OR "Industry 4" OR "fourth industrial revolution" OR "web of things" OR "Internet of Everything" OR "IoT" OR "CPS")

Databases: The databases Scopus and Web of Science were selected as the sources of primary studies because they group most of the Computer Science references including ACM, Elsevier, IEEE and Springer. Finally, inclusion/exclusion criteria and extraction data were set up.

Inclusion Criteria (IC):

- IC1: Studies published in the Computer Science and Engineering area
- IC2: Studies from journals, conferences or book chapters
- IC3: Studies published in English
- IC4: Studies concerning predictive quality
- IC5: Studies applying agents or data mining

Exclusion Criteria (EC):

- EC1: Editorials, books or erratums
- EC2: Studies not written in English
- EC3: Studies that were duplicated in the results
- EC4: Studies not fully available

Extraction Data: We planned to collect the following data for each paper: (i) goal; (ii) were there multi-agent systems? If so, what were the agent abilities applied?; (iii) was there data mining? If so, what was the mining approach used?; (iv) which quality characteristics were evaluated (effectiveness, efficiency, etc.)?; (v) which software measures took place? (vi) target audience.

3.2 Execution

On November 8th 2022, we ran the search string in the two search engines while automatically filtering IC1, IC2, IC3, EC1 and EC2. We got 684 publications (400 from Scopus and 284 from Web of Science), as presented in Fig. 1. After excluding duplicated studies (applying EC3, which resulted in 468 studies), we applied IC4 and IC5 selecting studies by analysing their titles (which resulted in 20 studies), and after that, their abstracts (which resulted in seven). When applying EC4, we got five publications total. Then, after reading the full publications and applying IC4 and IC5, it resulted in two studies for extracting data.

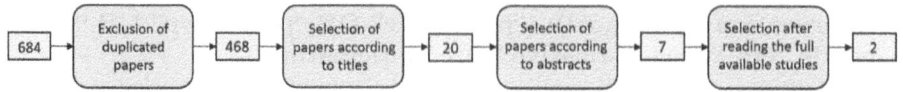

Fig. 1. Research steps and their respective results performed during the RR.

4 Results Analysis

Even though both of the analysed studies are related to a software system, neither had the end goal of evaluating it, and instead used it as means to an end (RQ1). One of the studies [8] aims at applying feature-learning methods to predict comfort and identify the usage context for products. To present their proposed architecture, the authors performed case studies using sensor data taken from a shoe with an integrated sensor. The other study [16] has the goal of predicting equipment component failure and equipment time to failure for both types of equipment involved: trucks and loaders. It proposes a data-driven solution to predict maintenance and improve the overall equipment effectiveness in mines.

Neither one of the studies use agents or multi-agent systems (RQ2), but both use data mining techniques to make predictions (RQ3): the two studies applied machine learning algorithms, where one considered decision trees [16] and the other neural networks and support vector machines (SVM) [8].

Regarding quality characteristics (RQ4), one study [16] evaluated efficiency, effectiveness and freedom from risk (safety risk mitigation), with the goal of predicting the devices maintenance and guaranteeing that the tool would be working safely. On the other hand, the other study [8] evaluated satisfaction (more specifically comfort, which is a subcategory of this quality characteristic) and context coverage, focusing on the context completeness part of this quality aspect. In Patil et al.'s study [16], the measures concern the prediction of success and not of quality: precision, recall, and false positive rate considering the events and detected failures. In Ghosh et al.'s study [8], for comfort prediction, a group of people wore a shoe with an integrated sensor while performing different tasks and answered surveys after. None of the studies applied published standards in their studies, such as ISO/IEC 25010 or other.

5 Final Remarks: Ongoing Works

This was the first step in our research where we tried to determine whether or not there had been any studies that aimed at predicting software quality using either multi-agent systems or data mining techniques. After performing such review and getting to two final studies, we conducted another review [3] focusing on evaluation methods for software quality.

From there, we designed an approach to predict the QinU in the context of smart environments based on agent mining while applying the ISO/IEC 25010 standard [11] as the main reference for the quality measures. For that, we will model and simulate the multi-agent system to represent both the actions of the human in such place as well as the smart environment around them and the expected behaviour from the devices according to the occupant's interactions. From there, we will mine the data produced from our simulation and classify it according to defined measures from ISO/IEC 25022.

In previous works, we used some measures to assess the quality of ubiquitous systems from the user's interaction with the software [6] and software tools that provide support to the QinU measurement process [18]. However, to our knowledge, there is no proposal for predicting the QinU without explicitly subjecting the user to the execution of the software, which could cause hazard to the person placed in the habitat, considering our goal is to measure the QinU for people with different degrees of motor impairments, either caused by ageing or some other reason.

Acknowledgements. This study has been supported by the Hauts-de-France Region and the REUNICE (Eunice Research) project funded by the European Union and the CNRS.

References

1. Abughazala, M.B., Moghaddam, M.T., Muccini, H., Vaidhyanathan, K.: Human behavior-oriented architectural design. In: Biffl, S., Navarro, E., Löwe, W., Sirjani, M., Mirandola, R., Weyns, D. (eds.) ECSA 2021. LNCS, vol. 12857, pp. 134–143. Springer, Cham (2021). https://doi.org/10.1007/978-3-030-86044-8_9
2. Albouys-Perrois, J., Sabouret, N., Haradji, Y., Schumann, M., Inard, C.: A co simulation of photovoltaic power generation and human activity for smart building energy management and energy sharing. In: V., C., E., F., A., G., F., P. (eds.) Building Simulation Conference Proceedings. vol. 4, pp. 2238 – 2245. International Building Performance Simulation Association (2019)
3. Angeloni, M.P.C., Oliveira, K.M., Grislin-Le Strugeon, E., Duque, R., Tirnauca, C.: A review on quality in use evaluation of smart environment applications: What's next? In: Companion of the ACM SIGCHI Symposium on Engineering Interactive Computing Systems (EICS '23 Companion), June 27–30, 2023, Swansea, United Kingdom (2023). https://doi.org/10.1145/3596454.3597177
4. Bonabeau, E.: Agent-based modeling: methods and techniques for simulating human systems. In: Proceedings of the National Academy of Sciences of the United States of America, vol. **99**, pp. 7280 – 7287 (2002). https://doi.org/10.1073/pnas.082080899
5. Cartaxo, B., Lima Pinto, G., Branco Soares, S.: Rapid reviews in software engineering. ArXiv **abs/2003.10006** (2020)
6. Carvalho, R., Andrade, R., Oliveira, K.: Aquarium - a suite of software measures for HCI quality evaluation of ubiquitous mobile applications. J. Syst. Softw. **136**, 101–136 (2018)
7. Felber, N.A., Tian, Y.J.A., Pageau, F., Elger, B.S., Wangmo, T.: Mapping ethical issues in the use of smart home health technologies to care for older persons: a systematic review. BMC Med. Ethics **24**(1) (2023).https://doi.org/10.1186/s12910-023-00898-w

8. Ghosh, D., Olewnik, A., Lewis, K.: Application of feature-learning methods toward product usage context identification and comfort prediction. J. Comput. Inf. Sci. Eng. **18**(1) (2018). https://doi.org/10.1115/1.4037435
9. Gramoli, L., Lacoche, J., Foulonneau, A., Gouranton, V., Arnaldi, B.: Needs model for an autonomous agent during long-term simulations. In: Proceedings of-4th IEEE International Conference on AI and VR, pp. 134–138 (2021). https://doi.org/10.1109/AIVR52153.2021.00031
10. Grislin–Le Strugeon, E., Oliveira, K.M., Zekri, D., Thilliez, M.: Agent mining approaches: an ontological view. Knowl. Eng. Rev. **36** (2021). https://doi.org/10.1017/S0269888921000114
11. ISO/IEC: 25010:2011. systems and software engineering — systems and software quality requirements and evaluation (square) — system and software quality models (2011). https://www.iso.org/standard/35733.html
12. Ketenci, U., Brémond, R., Auberlet, J., Grislin-Le Strugeon, E.: Drivers with limited perception: model and application to traffic simulation. Rech. Transp. Sécurité **2014**(1), 49–63 (2014)
13. Kitchenham, B., Charters, S.: Guidelines for performing systematic literature reviews in software engineering. Technical Report, EBSE 2007-001, Keele University and Durham University Joint Report (2007)
14. Malik, J., Azar, E., Mahdavi, A., Hong, T.: A level-of-details framework for representing occupant behavior in agent-based models. Autom. Construct. **139** (2022). https://doi.org/10.1016/j.autcon.2022.104290
15. Oladinrin, O.T., Mesthrige, J.W., Ojo, L.D., Alencastro, J., Rana, M.: Smart home technologies to facilitate ageing-in-place: professionals perception. Sustainability (Switzerland) **15**(8) (2023). https://doi.org/10.3390/su15086542
16. Patil, S.D., Mitra, A., Tuggali Katarikonda, K., Wansink, J.D.: Predictive asset availability optimization for underground trucks and loaders in the mining industry. Opsearch **58**(3), 751–772 (2021). https://doi.org/10.1007/s12597-020-00502-4
17. Petticrew, M., Roberts, H.: Systematic reviews in the social sciences: a practical guide. Blackwell Publishing Ltd, Oxford, UK (2006). https://doi.org/10.1016/B978-0-12-374708-2.00014-0
18. Salomon Garcia, S., Duque, R., Montaña, J., Tenes, L.: Towards automatic evaluation of the quality-in-use in context-aware software systems. J. Ambient Intell. Human. Comp. 1–26 (2022). https://doi.org/10.1007/s12652-021-03693-w
19. Tricco, A., Langlois, E., Straus, S., for Health Policy, A., Research, S., Organization, W.H.: Rapid reviews to strengthen health policy and systems: a practical guide. World Health Organization (2017)
20. Tricco, A., et al.: A scoping review on the conduct and reporting of scoping reviews. BMC Med. Res. Methodol. **16**(1), 1–10 (2016)
21. United Nations, D.O.E., Social Affairs, P.D.: Population 2030: demographic challenges and opportunities for sustainable development planning. Economical and Social Affairs (2015)
22. Wang, M., Hu, X.: Data assimilation in agent based simulation of smart environments using particle filters. Simul. Model. Pract. Theor. **56**, 36–54 (2015). https://doi.org/10.1016/j.simpat.2015.05.001
23. Wooldridge, M.: An Introduction to MultiAgent Systems, 2nd Edition. Wiley (2009)

A First Step Towards an Ecosystem Meta-model for Human-Centered Design in Case of Disabled Users

Christophe Kolski[1]([✉]) , Nadine Vigouroux[2] , Yohan Guerrier[1] ,
Frédéric Vella[2] , and Marine Guffroy[3]

[1] Univ. Polytechnique Hauts-de-France, CNRS, UMR 8201 – LAMIH, 59313 Valenciennes, France
{christophe.kolski,yohan.guerrier}@uphf.fr
[2] IRIT, UMR CNRS 5505, Université Toulouse 3, Toulouse, France
{nadine.vigouroux,frederic.vella}@irit.fr
[3] Wiztivi, Carquefou, France
marine.guffroy@wiztivi.com

Abstract. The involvement of the ecosystem or social environment of the disabled user is considered as very useful and even essential for the human-centered design of assistive technologies. In the era of model-based approaches, the modeling of the ecosystem is therefore to be considered. The first version of a meta-model of ecosystem is proposed. It is illustrated through a first case study. It concerns a project aiming at a communication aid for people with cerebral palsy. A conclusion and research perspectives end this paper.

Keywords: Ecosystem · Ecosystem model · Meta-model · User centered design · People with disabilities · Assistive technology · Cerebral palsy

1 Introduction

For many years, different models have progressively taken a more and more important place, and even become essential, in the field of Human-Computer Interaction, with the objective of designing interactive systems. It is possible to quote on this subject, without concern of exhaustiveness, but rather of representativeness: the task model [14], the user model [16], the architecture model [4], the context model [9, 13], the adaptation model [29] or the dialogue model [7], etc. For our part, our research concerns the field of disability, and in particular the human-centered design of assistive technologies.

According to the International Classification of Functioning, Disability and Health [31], the interactions between environmental factors (physical, social and attitudinal) and personal factors must be taken into account. Indeed, these factors influence the limitations of activity and social participation of people with disabilities for the design of assistive technologies. In this paper, we focus on the social environment of people with disabilities. This is referred to as the ecosystem by [25, 26]. This ecosystem is composed for example of family and/or professional caregivers, friends, therapists or colleagues, and this in relation to a set of activities of the person with a disability. Our motivations are the following: We propose that the ecosystem model is now one of the important models to consider. Such model could support a new generation of tools supporting the HCD (Human-Centered Design) process in disability domain. But in HCI and disability domains, there is a lack concerning ecosystem modeling. It is why we propose in this paper the first version of a meta-model of the ecosystem. As a meta-model, this one could be instanciated in the form of ecosystem models in projects.

After the presentation of the background, a first step towards a meta-model of the ecosystem is proposed. Then, we propose a first instantiation of the meta-model from a previous case study, using a reverse engineering approach. A discussion, a conclusion and research perspectives end the paper.

2 Background

This section concerns the implication of members of the ecosystem of disabled people in human-centered processes, with a view of ecosystem modelling and meta-modelling. Human-centered design (HCD) is based on ISO 9241-210 [28], which defines steps for its implementation: 1) Understanding and specifying the context of use; 2) Specifying user requirements; 3) Prototyping solutions to meet user needs; 4) Evaluating the solutions. In this approach, the end users, as well as various other stakeholders, are at the heart of the above-mentioned steps. However, these activities encounter implementation constraints for the design of assistive technologies dedicated to people with disabilities (motor impairments, cognitive impairments, communication disorders, etc.) [2]. Their participation in HCD activities may be differentiated according to their impairments. Many methods are available, allowing to involve them, with or without members of their ecosystem [5].

Many studies are interested in this question of the direct or indirect participation of ecosystem members of disabled people. Several participatory design approaches involve ecosystem members in the design, such as [17], or in both design and evaluation, for example [30]. In general, a so-called co-design approach can involve also members of the ecosystem, see for example [1]. Other authors position their approach more globally according to HCD and involve also members of the ecosystem in the design and evaluation, such as [3, 6, 8, 12]. For instance, in their study [12], Barros and colleagues report that key stakeholders (family and professional caregivers, physicians, physical therapists, and Parkinson's patients) were involved throughout the system design process of the target system; this one concerned the self-management of Parkinson's disease.

More generally, the literature shows that the ecosystem actors involved in human-centered design, participatory design, co-design…, are most often relatives (family and professional caregivers), associated therapists depending on the impairment (speech therapist, physiotherapist, occupational therapist, etc.) and people with professional expertise (doctor, specialized educators, teacher, etc.) depending on the assistive technology designed. Thus, there is a growing interest in the literature regarding the involvement of ecosystem members. Some papers suggest a first categorization of stakeholders in the projects. For example, Augusto and colleagues propose four types in their study [3]: Primary users (People with Down's Syndrome), Secondary users (Main carers), Tertiary users (Other system users), Other (Those interested in the system but no direct users). A global classification (represented as a tree) from a study concerning categories of medical device users is also available in [33], but it is not focused on disability domain. In general, we note that the papers available in the literature do not include a model or meta-model of the ecosystem.

In addition, it is important to note that several meta-models are available in HCI domain in general. They address various aspect of HCI design. They concern for instance context-aware adaptation of user interfaces [29], processes for highly supporting flexibility of user interfaces [10] or context-aware adaptation of mobile applications [15]. But, to our knowledge, none of them addresses more or less explicitly ecosystem considerations.

The involvement of the members of the ecosystem of the disabled person, in relation to the technical system, can be modeled according to the approach proposed below.

3 Proposal: Towards an Ecosystem Meta-model

We studied three user-centered design projects involving disabled people with communication disorders [23]. For each project, we analyzed their ecosystem. We found that the same types of actors could systematically be identified, and therefore involved in the methodological approach. Consequently, our aim was to propose a meta-model synthesizing the modeling elements involved, in relation to the three assistive technologies targeted [23, 26]. When considering other technologies, this meta-model could certainly be adapted or enriched.

As expressed by its *meta-* prefix, a meta-model (in our case, an ecosystem meta-model) is an abstraction allowing to describe models (i.e. ecosystem models) [18]. Figure 1 presents our first proposal for a meta-model aimed at providing support for the representation of the ecosystem of a user in a disability situation.

This meta-model is composed of the disabled person, one or more actors that can be considered as primary and secondary users of the assistive technology, and then one or more actors without interaction with the system (as defined in [2]). The disabled person has characteristics such as the type of impairment (e.g. mental or physical), the possible assistive technologies allowing to help them in different contexts of daily life, and finally the needs resulting from their disability situation. A primary actor (called Actor_P) is a person who directly uses the system. Most of the time, it is primarily the disabled person but other people in their environment can be primary actors of the system, in relation to one or more of its functionalities. For example a teacher who will regularly use a functionality of description of the day's activities of children with disabilities referring to this description on an interactive device [24] or an occupational therapist who design the assistive technology with and for the person with disabilities [8]. Concerning the secondary actor (called Actor_S) of the system, present in the meta-model, this user can interact for example, with the system if the primary user is unable to do so because of their disability or a constraining psychological situation such as stress. This actor also has a set of attributes. So, the type is intended to define whether the actor is, for example, a home care worker, a family member, a medical person, a teacher, etc. This person may also express needs in terms of communication assistance to help the person with a disability interact with people, known or unknown, in their environment.

The secondary actor has a direct relationship with the person with a disability, from a communication perspective. Finally, the proposed meta-model includes an actor (called Actor_N) who does not directly operate the system. This person is, for example, the recipient of messages formulated by the person with a disability. This actor is a person who does not necessarily know the primary user, and therefore may have difficulties to understand them. The disabled person interacts in this case with an actor of type Actor_N through the communication assistive technology, either by being assisted by an actor of type Actor_S, or with their assistive technology. This depends on the degree of the person's disability. In the following case study section, a first instantiation of the meta-model is provided and commented.

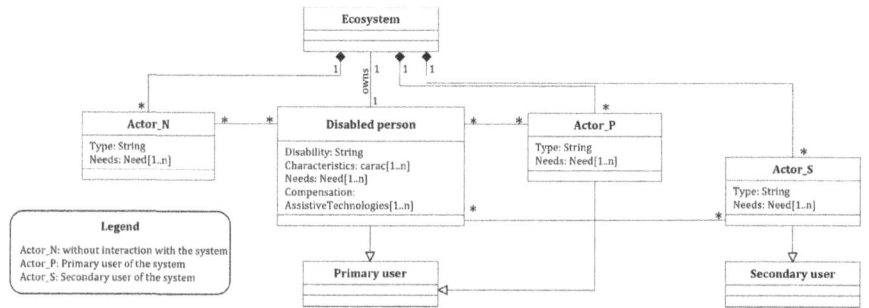

Fig. 1. Meta-model of ecosystem of user with disabilities.

4 First Case Study by Reverse Engineering

As described in Sect. 3, we studied three user-centered design projects involving disabled people and their ecosystem [23]. A meta-model resulting from this global analysis was proposed and described in Sect. 3. Using the principle of reverse engineering, it makes it possible to model the ecosystem a posteriori for each of these projects. This section illustrates a representative project.

This case study involved people with cerebral palsy. The project concerned the design, first prototyping and evaluation of a communication aid called ComMob (Communication and Mobility). After the description of the context, the ecosystem model related to this project is presented and commented. How the stakeholders in the ecosystem of the person with disabilities were involved in the HCD process is briefly presented. It is important to notice that the ecosystem model was built afterwards (by instantiation of the meta-model proposed), using a reverse engineering approach. It was possible to consult all the documents relating to the study and to interact with the members of the project team, with a focus on the role of the ecosystem in this project.

4.1 Context

In this case study, the target users have an athetosis-type cerebral palsy profile [32]. They make involuntary movements whose intensity can vary more or less strongly depending on the emotions. In addition, problems of precision in movements may make it difficult for them to use physical keyboards. These people also have speech problems due to dysarthria [11]. This means that the problems are usually in articulating words, but not in formulating correct sentences. Because of these different difficulties, people with this profile may require communication assistive technologies. A human-centered approach has been implemented with the aim of creating a communication assistive system for users with Cerebral Palsy [20]. In this system, called ComMob (Communication and Mobility), pictograms are organized in themes and categories. The user selects a set of pictograms to formulate sentences. The sentences are read by a text-to-speech system (see [19] for a description of ComMob).

4.2 Instantiation of the Ecosystem Meta-model

Figure 2 shows an instantiation of the meta-model proposed in Fig. 1, relative to this case study. The disabled person has cerebral palsy and is the primary user. This means that this person directly uses this Communication tool. Then we have a set of caregivers, both accustomed and unaccustomed to communicating with people of this profile. These categories are part of the ecosystem of the disabled person. The familiar caregivers do not need to use ComMob because they understand the disabled person. However, in the case that the user is under severe stress, cannot speak or move, the regular caregivers can use ComMob instead of the disabled person, or help their to use it. These people are either professionals or relatives. The second category (unaccustomed caregivers) needs the tool to communicate with the disabled person. Often they are professionals. Other occasional interlocutors belong to the category of secondary users. This means that they need ComMob to communicate with the disabled person. In this last category, we find in particular passers-by, salesmen, drivers or health professionals.

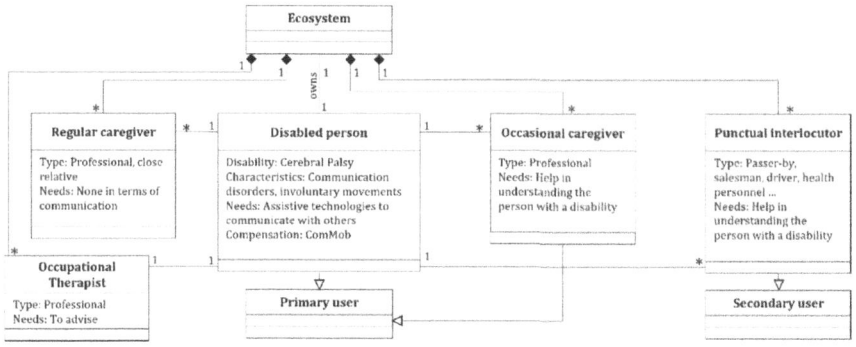

Fig. 2. Instantiation of the model in the case of ComMob project.

4.3 Actors' Involvement in the HCD Process

During the design of ComMob, the ecosystem was involved to gather the needs of future users. The disabled person, who is also the primary user, played a central role in the HCD process. Indeed, in the case of ComMob, the person with a disability was the designer of the application. Therefore, he had good knowledge in the field of cerebral palsy. In addition, he was in contact with an occupational therapist and other persons with the same disability. So he gathered information about the needs. A first mockup was made through his own needs, observations and testimonies obtained directly or on the internet. Then the designer conducted interviews with his family and his homecare providers to gather their opinions. Subsequently, tests were carried out with unaccustomed caregivers. For this purpose, the disabled person was in a public place. The task was to formulate a request to passers-by with ComMob [21]. Thanks to the data collected through observations and questionnaires filled in by passers-by, the project team was able to improve ComMob, for example by adding a button to repeat the previous sentence. Other tests were also performed in the laboratory. In one test, able-bodied participants used ComMob while simulating users with cerebral palsy, using a device manipulated by one of the evaluators. In this case, the ecosystem was not present, but an evaluator with cerebral palsy was an observer of the interactions performed by the participants [22].

5 Discussion

HCD has become a widely used design process over the years, resulting in a well-known standard. However, when the end users are disabled people with communication disorders, this model is no longer applicable and is no longer sufficient on its own. Indeed, the user cannot carry out alone the activities that the HCD provides. It is then necessary that actors of the ecosystem accompany the user. This paper proposes the first version of a meta-model to describe (by instantiation) such ecosystems.

The case study illustrates a first instantiation of this model, showing that these new actors have a place and a legitimate contribution in all the activities of the HCD. They have different professional expertise and social participation relationships with the disabled person. Furthermore, they may or may not be users of the assistive technology and may or may not assist the disabled person.

We note, however, that depending on the context in which the assistive technology will be used, the members of the ecosystem change. In a school context, the ecosystem will be composed of the educational team, while in a medical context it will be composed of health professionals. The implementation of HCD with the ecosystem requires some attention. The disabled user must not disappear from the design process [34], but must be involved by adapting the HCD tools or be represented by their ecosystem [27]. Indeed, the members of the ecosystem contribute and/or complete the collection of needs and accompany the user during the design and/or evaluation phases. The model may be very useful in highlighting the different actors involved in the project concerned (See also Sect. 6).

This ecosystem model, which takes the form of a UML class diagram, can be integrated into model-based approaches. Such a model can be linked to other models for design purposes. For example, through association or inheritance links with user models, it is possible to highlight or detail roles and characteristics. Connected to task models, it is possible to describe the tasks of each user category. Each context in which a user interacts with the assistive technology can be the subject of a context model. Each dialog associated with a human-machine interaction can be associated with a dialog model, and so on.

6 Conclusion

The software world is evolving quickly, with the advent of new model-based approaches. One observation is that models are playing an increasingly crucial role in projects. In parallel, the growing importance of the ecosystem in disability-related projects has led us to propose an ecosystem meta-model realized with the help of a UML class diagram. As a first validation, an illustration of the ecosystem model (again in the form of a class diagram), taking the form of instantiation of the meta-model, was provided. It comes from a case study concerning the ecosystem of people with cerebral palsy. Members of the ecosystem played important and diverse roles in the project.

The research perspectives are various. The proposal opens the way to the development of support tools for HCD approaches based on an ecosystem model or a library of reusable models. Moreover, it is necessary to analyze different projects, involving various disabilities, in order to highlight, for each of them, the corresponding ecosystem model. We have used the same modeling approach for other projects in which we have been involved; this work is not illustrated in this paper due to lack of space. We also plan to use it on new projects or projects where we have not been involved (provided the project documentation is available). The models obtained could lead to enrichments of the meta-model. As suggested in Sect. 5, we also plan to analyze the possible relationships between an ecosystem model (or the meta-model) and other types of models (user model, task model, context model, etc.). Another important perspective concerns the integration of the meta-model in model-driven engineering software environments and its implementation in various projects related to the disability domain.

The participation of the ecosystem in projects related to the field of disability is of growing importance. It should lead to the ecosystem model finding its place in the approaches and tools proposed by researchers and practitioners. The model could be associated with recommendations encouraging the systematic analysis of a set of possible categories of actor. It would support the study and choice of actors who could be involved in one or more phases of the user-centered design process [28]; for example, the family could be involved in the design and evaluation of mock-ups, while the doctor could be involved in the context understanding and specification phase, and so on. The model could also encourage reflection on the determination of primary (not necessarily only the disabled user) and secondary users. Indeed, depending on the characteristics and uses of the assistive technology, direct or indirect configuration or operation by an experienced or novice person may be necessary, depending on the context.

References

1. Ambe, A.M.H., Brereton, M., Soro, A., Buys, L., Roe, P.: The adventures of older authors: exploring futures through co-design fictions. In: Proceedings of CHI Conference on Human Factors in Computing Systems, CHI 2019, Glasgow, Scotland, UK, 04–09 May 2019. ACM (2019)

2. Antona, M., Ntoa, S., Adami, I., Stephanidis, C.: Chapter 15 - user requirements elicitation for universal access. In: Universal Access Handbook, pp. 15-1–15-14. CRC Press (2009)
3. Augusto, J., Kramer, D., Alegre, U., Covaci, A., Santokhee, A.: The user-centred intelligent environments development process as a guide to co-create smart technology for people with special needs. Univ. Access Inf. Soc. **17**(1), 115–130 (2018)
4. Bass, L., Little, R., Pellegrino, R., Reed, S.: The arch model: seeheim revisited. In: Proceedings of User Interface Developpers' Workshop, Seeheim (1991)
5. Beneteau, E.: Who are you asking? qualitative methods for involving aac users as primary research participants. In: CHI 2020, Honolulu, HI, USA, 25–30 April 2020 (2020)
6. Blanco, T., Berbegal, A., Blasco, R., Casas, R.: Xassess: crossdisciplinary framework in user-centred design of assistive products. J. Eng. Des. **27**(9), 636–664 (2016)
7. Bolchini, D., Paolini, P.: Interactive dialogue model: a design technique for multichannel applications. IEEE Trans. Multimedia **8**(3), 529–541 (2006)
8. Calmels, C., Mercardier, C., Vella, F., Serpa, A., Truillet, P., Vigouroux, N.: The ecosystem's involvement in the appropriation phase of assistive technology: choice and adjustment of interaction techniques. In: HCII 2021, Washington DC, USA, 24–29 July 2017 (2017)
9. Calvary, G., Coutaz, J., Thevenin, D., Limbourg, Q., Bouillon, L., Vanderdonckt, J.: A unifying reference framework for multi-target user interfaces. Interact. Comput. **15**(3), 289–308 (2003)
10. Céret, E., Dupuy-Chessa, S., Calvary, G.: M2FLEX: a process metamodel for flexibility at runtime. In: IEEE 7th International Conference on Research Challenges in Information Science, RCIS 2013, Paris, France, 29–31 May 2013, pp. 1–12 (2013)
11. Darley, F.L., Aronson, A.E., Brown, J.R.: Differential diagnostic patterns of dysarthria. J. Speech Hear. Res. **12**(2), 246–269 (1969)
12. De Barros, A. C., Cevada, J., Bayés, À., Alcaine, S., Mestre, B.: User-centred design of a mobile self-management solution for Parkinson's disease. In: Proceedings of 12th International Conference on Mobile and Ubiquitous Multimedia, Luleå, Sweden, December, pp. 1–10 (2013)
13. Dey, A.: Understanding and using context. Pers. Ubiq. Comp. **5**, 4–7 (2021)
14. Diaper, D., Stanton, N.A.: The Handbook of Task Analysis for Human-Computer Interaction. Lawrence Erlbaum Associates, Mahwah (2004)
15. de Farias, C.R.G., Leite, M.M., Calvi, C.Z., Pessoa, R.M., Pereira Filho, J.G.: A MOF meta-model for the development of context-aware mobile applications. In: Proceedings of the 2007 ACM Symposium on Applied Computing, pp. 947–952. ACM (2007)
16. Fischer, G.: User modeling in human-computer interaction. User Model. User-Adap. Inter. **11**(1–1), 65–86 (2001)
17. Frauenberger, C., Makhaeva, J., Spiel, K.: Blending methods: developing participatory design sessions for autistic children. In: Proceedings of Conference on Interaction Design and Children, Stanford, CA, USA, 27–30 June 2017, pp. 39–49. ACM (2017)
18. Gonzalez-Perez, C., Henderson-Sellers, B.: Metamodelling for Software Engineering. Wiley, Chichester (2008)
19. Guerrier, Y.: Proposal of a software support for information entry in degraded situations: application for users with Cerebral Palsy athetosis in transport and daily activities (in french), PhD Thesis, Univ. Valenciennes, France (2015)
20. Guerrier, Y., Kolski, C., Poirier, F.: Towards a communication system for people with athetoid cerebral palsy. In: Kotzé, P., Marsden, G., Lindgaard, G., Wesson, J., Winckler, M. (eds.) INTERACT 2013. LNCS, vol. 8120, pp. 681–688. Springer, Heidelberg (2013). https://doi.org/10.1007/978-3-642-40498-6_61
21. Guerrier, Y., Naveteur, J., Kolski, C., Poirier, F.: Communication System for Persons with Cerebral Palsy. In: Miesenberger, K., Fels, D., Archambault, D., Peňáz, P., Zagler, W. (eds.) ICCHP 2014. LNCS, vol. 8547, pp. 419–426. Springer, Cham (2014). https://doi.org/10.1007/978-3-319-08596-8_64

22. Guerrier, Y., Naveteur, J., Kolski, C., Anceaux, F.: Discount evaluation of preliminary versions of systems dedicated to users with cerebral palsy: simulation of involuntary movements in non-disabled participants. In: Antona, M., Stephanidis, C. (eds.) HCII 2021. LNCS, vol. 12768, pp. 71–88. Springer, Cham (2021). https://doi.org/10.1007/978-3-030-78092-0_5
23. Guerrier, Y., Vigouroux, N., Kolski, C., Vella, F., Guffroy, M., Teutsch, P.: Conception centrée utilisateur d'aides techniques pour des utilisateurs en situation de handicap avec troubles de la communication : retour d'expérience pour une participation systématique de leur écosystème. Revue des Interactions Humaines Médiatisées **21**(1), 29–56 (2020)
24. Guffroy, M.: Adaptation of evaluation methods in the design of a digital application for a young audience with autism spectrum disorders (in french). PhD Thesis, Université du Maine, France (2017)
25. Guffroy, M., Guerrier, Y., Kolski, C., Vigouroux, N., Vella, F., Teutsch, P.: Adaptation of user-centered design approaches to abilities of people with disabilities. In: Miesenberger, K., Kouroupetroglou, G. (eds.) ICCHP 2018. LNCS, vol. 10896, pp. 462–465. Springer, Cham (2018). https://doi.org/10.1007/978-3-319-94277-3_71
26. Guffroy, M., Vigouroux, N., Kolski, C., Vella, F., Teutsch, P.: From human-centered design to disabled user & ecosystem centered design in case of assistive interactive systems. Int. J. Sociotechnol. Knowl. Dev. **9**(4), 28–42 (2017)
27. Hendriks, N., Slegers, K., Duysburgh. P.: Codesign with people living with cognitive or sensory impairments: a case for method stories and uniqueness. CoDesign **11**(1), 70–82 (2015)
28. ISO: Ergonomics of human-system interaction—Part 210: Human-centred design for interactive systems. ISO 9241-210:2019, ISO, Geneva (2019)
29. Motti, V.G., Vanderdonckt, J.: A computational framework for context-aware adaptation of user interfaces. In: Proceedings of RCIS 2013, Paris, France, 1–12 May 2013. IEEE (2013)
30. Nasr, N., et al.: The experience of living with stroke and using technology: opportunities to engage and co-design with end users. Disab. Rehab. Assistive Tech. **11**(8), 653–660 (2016)
31. OMS: International Classification of Functioning, Disability and Health (ICF), Fifty-fourth World Health Assembly on 22 May 2001 (2001)
32. Rosenbaum, P., et al.: A report: the definition and classification of cerebral palsy April 2006. Dev. Med. Child Neurol. Suppl. **109**(no suppl 109), 8–14 (2007)
33. Shah, S.G.S., Robinson, I.: Medical device technologies: who is the user? Int. J. Healthc. Technol. Manag. **9**(2), 181–197 (2008)
34. Wobbrock, J.O., Gajos, K.Z., Kane, S.K., Vanderheiden, G.C.: Ability-based design. Commun. ACM. ACM **61**(6), 62–67 (2018)

Evaluation of a Social Robot System for Performance-Oriented Stroke Therapy

Alexandru Umlauft(✉) [ID] and Peter Forbrig [ID]

Universität Rostock, Fakultät Für Informatik, Lehrstuhl Für Softwaretechnik,
Albert-Einstein-Str. 22, 18059 Rostock, Germany
{alexandru-nicolae.umlauft,peter.forbrig}@uni-rostock.de

Abstract. Over the past few years, social robots have found utility in rehabilitative therapies across various fields. However, most of these studies did not incorporate extended clinical trials or focus on enhancing the health status of the patients with a medical stroke therapy. In this study, we present a research project in which a social robot serves as an instructor for stroke patients during their rehabilitation sessions. In this paper we briefly present our social robot application and focus on its evaluation. We provide an overview of relevant previous work and outline the system setup and experimental study. Lastly, we discuss our evaluation and conclude our findings.

Keywords: Social Robot · Stroke · Therapy · Technology Acceptance · Usability

1 Introduction

The world is facing a shortage of medical staff in healthcare systems across the globe. Factors such as population growth, aging demographics, and increased healthcare demands have contributed to this crisis. As a result, finding innovative solutions to bridge this gap and enhance patient care has become imperative. One promising solution lies in the use of social robots in rehabilitation. While they cannot replace the expertise and compassion of human healthcare providers, social robots can act as valuable tools to supplement and support their work.

In our works, we focus on stroke therapies together with a socially assistive robot (SAR). The work was done inside the E-BRAiN project. The robot is not applying any physical help and instead only instructs and provides feedback. This paper describes our ongoing efforts of building an effective SAR system for stroke rehabilitation and current results of the evaluation. The biggest contribution to the field of HRI lies in the preliminary evaluation of our long-term SAR study with a technology acceptance questionnaire (TAM) with the ALMERE-model in combination with a usability interview. These interviews aimed to identify good and improvable aspects of the interaction design of social robots in the environment of stroke rehabilitation.

In the following we provide a brief overview of similar SAR systems and describe the implemented stroke therapies and system overview. Then we explain the approach we used for our implementation and present the results. Finally, we discuss the findings

---

© The Author(s), under exclusive license to Springer Nature Switzerland AG 2024
M. Harrison et al. (Eds.): EICS 2023 Workshops, LNCS 14517, pp. 20–27, 2024.
https://doi.org/10.1007/978-3-031-59235-5_3

of the technology acceptance questionnaire and give a broad overview of the usability results.

2 Related Works

In this section we focus on the evaluation of social robots. To evaluate the usability SAR system, many methods with different goals can be deployed. According to the review done by Jung et al. [9] in 2021, 62% of observed evaluations use questionnaires, at the second place video analysis are implemented with 21% of the share and interviews with the users have a share of 13%. They also state, that depending on the concerning three dimensions of the attitude towards a robot "utilitarian" and "hedonic" aspects and "trust", a reasonable combination of evaluation techniques should be used. Questionnaires and interviews can cover all these dimensions. In these dimensions, constructs like such of the technology acceptance and "emotional attractiveness" are included.

Technology acceptance aims to predict, if a new technology will later be used. For this, typically a questionnaire will be filled out. From these answers, certain constructs are supposed to deduct the probability of an intention to use. The original TAM used the constructs "perceived ease of use" and "perceived usefulness".

Heerink et al. [8] presented the ALMERE model. It was developed specifically to work with elderly persons to test their acceptance for social robots. It is an enhanced model from the UTAUT- technology acceptance model [14] hereby the UTAUT model originates from the first TAM- model [2, 3]. In their tests, they employed and tested it on 4 applications. It showed good and robust results of predicting the intention to use and the model has been later employed by others later as well.

Feingold-Polak [5, 12] present a stroke rehabilitation system with different instructor options in their work. In their experiment, they conduct a long-term study with a 5–7 week study. Each patient performs 15 sessions with the social robot pepper, a digital avatar and only written manuals/instructions. One of the goals was to observe and compare how well the different solutions are perceived. The tasks were based on "serious games" and ranged from simple exercises such as placing cups, playing cards or tidying up a shelf. For the evaluation, they used a modified questionnaire to test the usability with direct questions like "Was the information provided by the system clear to you?" to answer on a 5-point scale and open-ended questions like e.g. "What would you add or change?". In their end, the social robot was rated slightly better than the other options.

In a different work of their group done by Koren et al. [10], they conducted qualitative, semi-structured interviews to gain more insights about trust in social robot rehabilitation systems. One finding ins that work was, that out of the patients who received sessions with the SAR system, 78% of them were satisfied with the therapeutic procedure.

Lee et al. [10] displayed a SAR system for stroke rehabilitation with a focus on the recognition of movement during the exercises. For the development of their system they interviewed 3 therapists for understanding how the interaction should be modelled. For their experiments they observed 3 movement exercises which are done sitting. The system aims to provide precise feedback. For their experiments they recruited only healthy participants. Their evaluation is regarding the precision and usability of the movement recognition and the robot's feedback. They employed a usability questionnaire originally from Fasola et al. [4], which again targets multiple evaluation points. It uses questions such as "While you were exercising with [the robot], how much did you feel as if you were interacting with an intelligent being?" to determine the construct "Social presence". Their usability results state, that the participants rather agree to the questions, if the robot is perceived as useful.

For our work, we chose the ALMERE model as presented from Heerink et al. [8], for the assessment of the technology acceptance.

To engage the question of usability from a slightly different angle, we also aim to gain more insights from the patients about our system. For this, we aimed for usability interviews as comparably proposed by Nielsen [11]. Such usability-studies target to identify which specific features and designs were helpful (or not) to include in a future robot system iteration. For the usability we chose the approach of a pluralistic walkthrough [11] and asked the patients to "think loud".

3 Our System

We have described the architecture and setup in earlier works [6, 7]. Our experiments utilize the robot Pepper as depicted in Fig. 2. We have integrated an android tablet, a touch panel for confirmation, a therapy administration computer. The computer unit also controls the interaction and the actual content of the robot's feedback. The system receives explicit confirmation through the physical touch panel and, in certain therapies, speech input from the patient. All components communicate with each other using the MQTT protocol (Fig. 1).

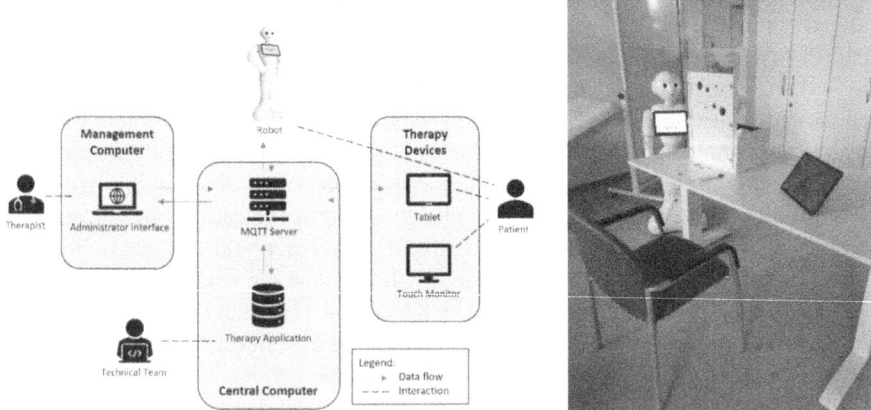

Fig. 1. Overview of the system components   **Fig. 2.** The visible patient setup

We have developed a website interface for managing the therapy sessions and patients. During the therapy session, both the tablet and the robot display the images, videos, and instructional text on their screens. Patients confirm certain actions by touching elements on the screen.

From the patient's perspective, at the beginning of a therapy session, they will sit down and interact with the nearby tablet or monitor to initiate the therapy.

The interaction is modeled with a finite state machine similar and we presented our intermediate results in an earlier work [1]. The medical team described the therapy they wanted as relatively static and thus, the therapy text and speech content is not dynamically created. The decisions of the robot should always be understandable. In addition, for ethical reasons, no machine learning methods should be included in order not to cause probability-based failures or problems for affected patients. The interaction logic was mainly implemented in Python and the end devices (tablet, pepper robot) required Android apps.

4 Ongoing Therapy Study

4.1 Implemented Therapies

We developed four therapies, presented in previous works [1, 6, 7]. Arm ability training (AAT) targets minor difficulties in arm movement, mirror therapy (MT) focuses on arm paralysis primarily affecting mostly one side, and Arm basis training (ABT) is employed for severe arm paralysis. Additionally, neurovisual therapy is provided for patients experiencing neglect-related impairments. Neglect-related impairments often results in a limited field of vision for the patient. Varying from therapy and amount of sessions, the robot sessions take between 50–70 min, and during the first session (with longer dialogues) around 70–90 min.

4.2 Medical Study Design

The study follows a crossover design and primarily pursues the medical goal of comparing the effectiveness of rehabilitation between human therapy and the robotic system.

For this purpose, patients were recruited from the area surrounding our study site (Greifswald, Germany) and included in a four-week study. Before starting, potential patients are screened. Our study selection screened for patients who have reasonably high chances of recovery. Patients are typically experiencing other therapies at this point. Of the 4 weeks of our study, they will be treated with our robotic system for 2 weeks. The other 2 weeks they will continue their other regular therapy. Assessments are carried out with the patients at various times during the 4 weeks. These include e.g. the Fugl-Meyer test [13]. Based on certain of these values, it is then considered how much patients have improved.

5 Evaluation

Here we describe our approaches for the evaluation of our system. We were able to talk with 6 patients in total so far. We aimed for a two-sided evaluation session by using a technology-acceptance approach and a usability analysis. For the latter, the authors try to summarize the experiences of the usability interview objectively.

Each evaluation session started after the regular robot therapy session of that day. This was requested and planned by the medical team to easy for the patient and to prioritize the medical study. At the beginning of each evaluation session, we started to hand out the questionnaires for the technology acceptance. In most cases, patients were not able to write with a pen properly, thus the evaluator went with the patient through the questions and wrote the answers down.

For the usability interview, we proceeded to start the regular therapy, whereby this should only help patients to better focus on the usability aspects and they should not exercise. In general, the session walkthrough started from the beginning and ends at the beginning of the second exercise. At certain points, we included breaks. The evaluation sessions lasted between 30–45 min.

5.1 Preliminary Results Technology Acceptance Modelling with the ALMERE Model

The results are displayed in Fig. 3. Overall most constructs were positively answered and anxiety low.

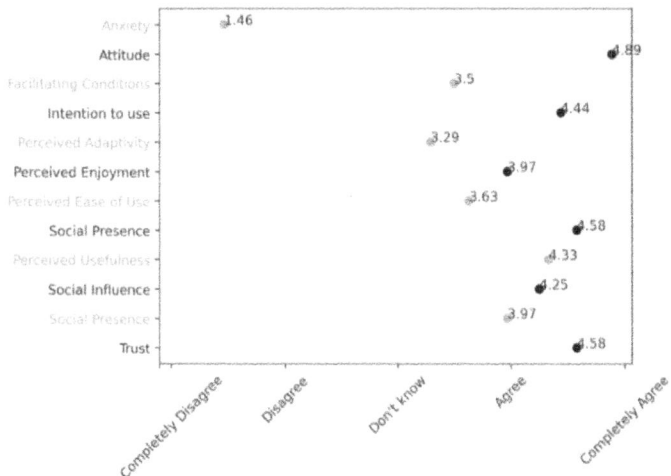

Fig. 3. The results of the ALMERE technology acceptance questionnaire. For easier understanding we added numbers displaying the exact numbers of the results 1.00 = "completely disagree" and 5.00 = "completely agree".

The results concerning the construct of anxiety with 1.46 are rather low, which can be perceived as a good sign for the later usage of the system, since this indicates that the patients are not anxious to work with the robot. All other constructs are on average rather positively answered.

5.2 Preliminary Usability Interview Experiences

Overall, the 6 interviewed patients appeared to like the system and the interaction. We went through the interviews and there were no relevant complains on parts of the therapy system. All patients seemed to be satisfied with the system.

However, most patients did not work with the style of "thinking loud" and instead, the interviewer had often to give prompts. In these cases, patients were asked what they thought about e.g. certain robot interactions or layout design. Thus, some parts of the interview became comparable to a semi-structured interview.

Most of the time, patients usually replied to a question like "What do you think of this [point]?" with e.g. "It's good." and did not have a wish to make things differently.

We had two cases, where we recorded change requests. One patient stated that "the robot is too slow". This particular patient was doing the neurovisual therapy and meant this way, that the layout-elements and touch-interaction (searching and touching objects on the screen as a training to learn a search pattern) was not reactive enough. We should mention here, that this point was only mentioned by this patient, and the other 5 were fine with the interaction speed.

The other case was, one patient stated, that during the exercise explanations with video and image displays, he does not feel connected to the shown materials. The point was, that our media files are featuring the same 2 women demonstrating the exercises of our therapies. Naturally their movements might differ and their arms and hands appear thinner, than that of a man.

6 Discussion and Limitation

Social robots have been academically tested in different scenarios concerning medical applications. But often times, detailed evaluation about the interaction design and acceptance are often lacking.

Here we presented a social robot system for stroke rehabilitation and showed the initial results of our evaluation. We applied a technology acceptance questionnaire with the ALMERE model and did "thinking loud" usability interviews with the patients. Overall our system, seems to be well accepted and only few change wishes were stated. In comparison with the evaluation of Feingold-Polak et al. [5] our results seem to be similar. The patient did not show signs of anxiety against the robots and generally the patients seemed to well accept the new technology. Their patients seemed also to be happy with the robot system.

It is rather surprising, that patients were not criticizing the system more, as the system interaction might appear as rather dull and monotone. Of course, the rehabilitation tasks are the most important and time-filling part of a session and because of this will keep the patient busy. Yet we thought that these parts where the robot talks for several minutes would have been encountered.

The mentioned case from one patient where the system interaction was too slow for the patient, could be a unusual case. During the interview we learnt that when the patient is not lying down, the patient feels pain. In order to distract from the pain, the patient tries to do the exercises fast and potentially sooner return to a more pain-free position.

Our system, so far is designed to let the robot fully speak the text during the exercises and to nudge patients into learning the search pattern strategy with their eyes. We had the balancing act to import a medical therapy into a version with a social robot, which naturally bring some technical disadvantages, such as planned delays. Yet we think that a system needs to adapt to the patient's needs or wishes, but a rehabilitation system still needs to actually provide the stimulation to rehabilitate.

The main limitations of our evaluation are that we were only able to find 6 patients for the evaluation so far and furthermore that only a selected few of them actually worked with the usability interview to "think loud". It might be a good result that our system received good feedback, but from earlier works with new technologies where users do not have comparisons to other technologies, the answers seem to be often positively skewed.

Additionally, during the usability interviews the patients appeared to be very motivated to work with a robot system and this might have added to the strong positive results. Often times, patients stated something like "I wouldn't be here, if I would not believe that it [the robot system] helps me".

7 Conclusion

We displayed our current works of a SAR system for stroke therapies with medical therapies. We contribute to the workshop by presenting the results of our current evaluation of such a system and our experiences how the patients thought about the system. Our robot system showed positive results in the technological acceptance questionnaires and patients did not have complaints. While it appears, that the usability of our system has been positively received, we think that there still enough "edge cases" which we didn't found yet. With time and technological advances, social robots might become more common and patients may gain more reference knowledge, what they would like to experience in a therapy setting.

We are still recruiting the patients for the medical study. Based on our results, we will develop new version of our system. In the future, we hope to continue work with multiple cities and different medical teams to test a social robot therapy system on a larger scale.

Acknowledgements. This joint research project "E-BRAiN - Evidence-based Robot Assistance in Neurorehabilitation" is supported by the European Social Fund (ESF), reference: ESF/14-BM-A55-0001/19-A01, and the Ministry of Education, Science and Culture of Mecklenburg-Vorpommern, Germany. The sponsors had no role in the decision to publish or any content of the publication.

References

1. Bundea, A., Bader, S., Forbrig, P.: Interaction and dialogue design of a humanoid social robot in an analogue neurorehabilitation application. In: Zimmermann, A., Howlett, R.J., Jain, L.C., Schmidt, R. (eds.) KES-HCIS 2021. Smart Innovation, Systems and Technologies, vol. 244, pp. 76–85. Springer, Singapore (2021). https://doi.org/10.1007/978-981-16-3264-8_8

2. Davis, F.D.: Perceived usefulness, perceived ease of use, and user acceptance of information technology. MIS Q. **13**(3), 319 (1989). https://doi.org/10.2307/249008
3. Davis, F.D., Bagozzi, R.P., Warshaw, P.R.: User acceptance of computer technology: a comparison of two theoretical models. Manag. Sci. **35**(8), 982–1003 (1989). https://doi.org/10.1287/mnsc.35.8.982
4. Fasola, J., Mataric, M.: A socially assistive robot exercise coach for the elderly. J. Hum.-Robot Interact. **2**, 2 (2013). https://doi.org/10.5898/JHRI.2.2.Fasola
5. Feingold-Polak, R., Barzel, O., Levy-Tzedek, S.: A robot goes to rehab: a novel gamified system for long-term stroke rehabilitation using a socially assistive robot-methodology and usability testing. J. Neuroeng. Rehabil. **18**(1), 122 (2021). https://doi.org/10.1186/s12984-021-00915-2
6. Forbrig, P., Bundea, A., Bader, S.: Engineering the interaction of a humanoid robot pepper with post-stroke patients during training tasks. In: Companion of the 2021 ACM SIGCHI Symposium on Engineering Interactive Computing Systems, pp. 38–43. ACM, New York (2021). https://doi.org/10.1145/3459926.3464756
7. Forbrig, P., Bundea, A., Pedersen, A., Platz, T.: Using a humanoid robot to assist post-stroke patients with standardized neurorehabilitation therapy. In: Nagar, A.K., Jat, D.S., Marín-Raventós, G., Mishra, D.K. (eds.) Intelligent Sustainable Systems. Lecture Notes in Networks and Systems, vol. 334, pp. 19–28. Springer, Singapore (2022). https://doi.org/10.1007/978-981-16-6369-7_3
8. Heerink, M., Kröse, B., Evers, V., Wielinga, B.: Assessing acceptance of assistive social agent technology by older adults: the Almere model. Int. J. Soc. Robot. **2**(4), 361–375 (2010). https://doi.org/10.1007/s12369-010-0068-5
9. Jung, M., Lazaro, M.J.S., Yun, M.H.: Evaluation of methodologies and measures on the usability of social robots: a systematic review. Appl. Sci. **11**(4), 1388 (2021). https://doi.org/10.3390/app11041388
10. Koren, Y., Polak, R.F., Levy-Tzedek, S.: Extended interviews with stroke patients over a long-term rehabilitation using human-robot or human-computer interactions. Int. J. Soc. Robot. **14**(8), 1893–1911 (2022). https://doi.org/10.1007/s12369-022-00909-7
11. Nielsen, J.: Usability Engineering. AP Professional, Cambridge (1993)
12. Polak, R.F., Levy-Tzedek, S.: A Robot Goes to Rehab: A Novel Gamified System for Stroke Rehabilitation using a Socially Assistive Robot: Methodology & Feasibility Testing (2020)
13. Singer, B., Garcia-Vega, J.: The Fugl-Meyer upper extremity scale. J. Physiother. **63**(1), 53 (2017). https://doi.org/10.1016/j.jphys.2016.08.010
14. Venkatesh, V., Morris, M., Davis, G., Gordon, F.: User acceptance of information technology: toward a unified view. MIS Q. **27**(3), 425 (2003). https://doi.org/10.2307/30036540

MUMR-MIODMIT: A Generic Architecture Extending Standard Interactive Systems Architecture to Address Engineering Issues for Rehabilitation

Axel Carayon, Célia Martinie(✉), and Philippe Palanque

IRIT, University Toulouse III – Paul Sabatier, Toulouse, France
{axel.carayon,celia.martinie,philippe.palanque}@irit.fr

Abstract. Several technologies and tools have been proposed to assist rehabilitation team in performing exercises. They can help rehabilitation specialists to manage more patients when practicing their exercises, either in presence or remotely. They also make it possible to support patients to do parts of the program at home. Few contributions deal with supporting to the engineering of training platforms for post-stroke rehabilitation. Designers and developers of training platforms for post-stroke rehabilitation do not have a conceptual framework to rely on to develop training Hardware (HW)/Software (SW) platforms. Furthermore, the development and implementation of training platforms for post-stroke rehabilitation requires being able to engineer training platforms with multiple HW and SW components and to engineer coordinated and relevant connections between these components. This paper describes the MUMR (Multi User Multi Role)-MIODMIT technique to design the architecture of the interactive systems for post-stroke rehabilitation training. This technique is based on the MIODMIT description technique for the architecture of interactive systems and enables to describe the main aspects of a post-stroke rehabilitation training platform. The proposed technique is applied to two examples of rehabilitation training platforms.

Keywords: Interactive Systems · Software Architecture · Rehabilitation

1 Introduction

According to the World Stroke Organization in the Global Stroke Factsheet released in 2022 [23], lifetime risk of developing a stroke has increased by 50% over the last 17 years and now, 1 in 4 people is estimated to have a stroke in their lifetime. A stroke may cause several types of impairments. The aim of a stroke rehabilitation program is to reduce the after-effects of stroke and restore the capacity of stroke victims. Post-stroke or post-brain injuries rehabilitation is a long-term process defined by therapists and tuned to the specific damages and their consequences for each patient. This process requires the definition of well-defined multiple exercises to be performed multiple times by each patient over a long period of time. Several technologies and tools have been proposed to assist rehabilitation team in performing exercises. They can help rehabilitation specialists to manage more patients when practicing their exercises, either in presence or remotely. They also make it possible to support patients to do parts of the program at home.

Until today, few contributions deal with supporting to the engineering of training platforms for post-stroke rehabilitation. Designers and developers of training platforms for post-stroke rehabilitation do not have a conceptual framework to rely to develop training HW/SW (HardWare/SoftWare) platforms. Furthermore, the development and implementation of training platforms for post-stroke rehabilitation requires being able to engineer training platforms with multiple HW and SW components and to engineer coordinated and relevant connections between these components. We thus propose a the MUMR (Multi User Multi Role) – MIODMIT technique to design the architecture of the interactive systems supporting post-stroke rehabilitation training. This technique is based on the MIODMIT description technique for the architecture of interactive systems [6] and enables to describe the main aspects of a post-stroke rehabilitation training platform.

The paper is structured in the following way. Section 2 presents the needs for support to the design of the architecture of Interactive systems for rehabilitation training. Section 3 highlights the main principles of the MIODMIT architecture. Section 4 presents the extensions proposed to the MIODMIT architecture. Section 5 illustrates the proposed technique on two examples of rehabilitation training platforms. Section 6 concludes the paper.

2 Need for Support to the Design of the Architecture of Interactive Systems for Post-stroke Rehabilitation Training

Interactive systems for post-stroke rehabilitation training aim to support rehabilitation specialists in conducting the recovery of their patients. These training systems automate specific training tasks previously directed by rehabilitation experts. They support patient guidance through selected recovery exercises for motor, proprioceptive, sensory, cognitive, sensory and cognitive abilities.

2.1 Needs and Goals in Rehabilitation Training

Stroke remains the third leading cause of death and disability combined [21], meaning that health institutions have a lot of patients to take care of and to help for their rehabilitation. Interactive systems for post-stroke rehabilitation training have several advantages. They can help rehabilitation specialists to manage more patients in their practice or remotely. In addition, they can provide support for patients to do parts of the program at home.

As stated in [13], the main principles of a post-stroke rehabilitation program are:

- Goal setting: establishment of specific, measurable, and time-dependent recovery goals to guide management.
- High-intensity practice: increased therapy or intervention.
- Multidisciplinary team care: a team of medical, nursing, therapy and social-work staff who provide rehabilitation input and coordinate their work with regular meetings.

- Task-specific training: rehabilitation approaches where specific functional tasks are practiced repeatedly.

Recent work from Forbrig et al. [17] advocates for human-centered design of neuro-rehabilitation software and proposes to analyze rehabilitation team needs and context of use.

In the scientific literature, robots and robotic devices are widely represented because they provide several benefits such as automatizing therapy procedures [18] but they tend to be complex, big, expensive and hard to operate. Because of that, in the case of home rehabilitation, which has been proven to be effective [14], we need to rely on cheaper, easy to use devices that the user can use on its own without the need of a therapist. Furthermore, post-stroke rehabilitation involves several roles (patients, therapists of different disciplines...) [5], requires to support the integration of multiple and evolving rehabilitation goals and tasks, as well as to connect high-level goals with low-level exercise tasks [5], and such integration requires to be able to integrate several hardware and software technologies [5].

Thus, all of these require to be able to represent different systems with different type of users, in different environments.

If we summarize all the previous principles, guidelines and recommendations, in order to be exhaustive, an architecture for interactive post-stroke rehabilitation systems must:

- Represent complex systems that can use multiples technologies or devices, combining several input or outputs
- Takes into account that several users, that have a different role, can be interacting with the same system
- Be able to represent complex interactions combining multiple input or outputs
- Represent the interaction between multiple systems and/or users in a synchronous or asynchronous manner

2.2 Related Work in the Architecture of Interactive Systems

While the Seeheim model [17] was the first known architectural model for interactive systems (and who clearly distinguish the concepts of presentation, dialog and interface with the functional core), it is a very abstract model that do not describe with enough details the interactions between the many parts of the interactive systems. From this first architectural model emerged the ARCH model [2], who refined the presentation part into a physical interaction part and a logical interaction part.

Since then more architectures have emerged, such as an architecture that identify the various components of multimodal car systems [10], a multi-agent based architecture for decentralized CSCW interactive systems [12], or an architecture focusing on a specific kind of interaction such as speech interaction [11]. The work we present is based on the MIODMIT [6] architectural model, one of the latest attempts to propose a generic way to describe any kind of interactive systems. For more details, we invite the interested reader to read its related work section, as it provides a quite accurate and complete description of the current state of the art.

3 Interactive Systems Architecture: Summary of H-MIODMIT

Figure 1 presents an architectural view (from left to right) of the operator, the interactive command and control system, and the underlying cyber-physical system (e.g., an aircraft engine). This architecture is named H-MIODMIT (Human-Multiple Input and Output Devices and Multiple Interaction Techniques) [4] and is a simplified and extended version, for the human part, of the MIODMIT generic architecture for multi-modal interactive systems [6]. Rounded boxes represent functions and arrows represent information flows. The top part of the figure differentiates the hardware and software components of an interactive system. Interaction with the operator takes place only through manipulation of hardware input devices and perception of information from hardware output devices (which includes the cyber-physical system).

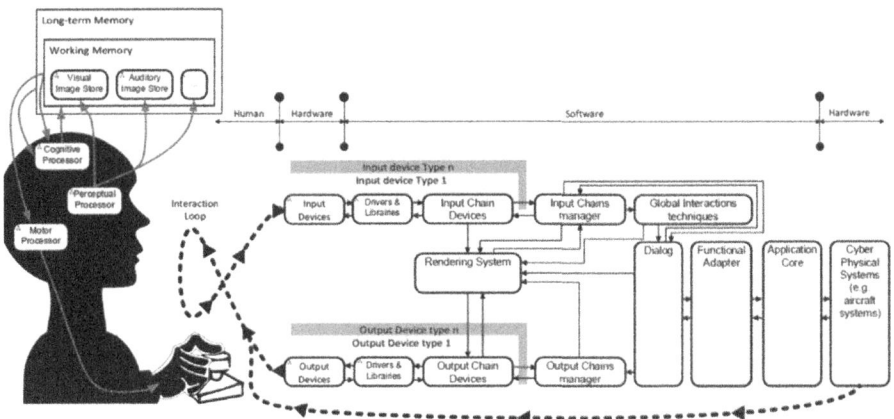

Fig. 1. H-MIODMIT architecture (adapted from [4])

The System Side. The right side of the Software section of the architecture (starting with Dialog box) corresponds to what is usually called interactive applications. This is where HCI methods such as task analysis are needed for building usable application that fit the operators' work [7]. In the context of command and control systems, this interactive application is connected to the so called Cyber-Physical System to operate (which is also most of the time composed of hardware and software components).

The Human Side. The left side of Fig. 1 represents the operator's side. The drawing is based on work that models the human as an information processor [1], based on previous research in psychology. In that model, the human is presented as a system composed of three interconnected processors. The perceptive system senses information from the environment – primarily the visual, auditory and tactile systems as these are the most commonly used for interacting with computers. The motor system allows operators to act on the real world. Target selection (a key interaction mechanism) has been deeply studied [20]; for example, Fitts' Law provides a formula for predicting the difficulty for an operator to select a target, based on its size and distance [8]. The cognitive system is in charge of processing information gathered by the perceptual system, storing that information in memory, analyzing the information and deciding on actions performed using the motor system. The sequential (sometimes parallel) use of these systems (perceptive, cognitive and motoric) while interacting with computers is called the Human-Computer Interaction Loop (HCIL). The operator's memory (decomposed into short term and long-term memory) is known for being subject to alteration and decay [15].

The Specificities of the Interaction. The top left of the Software section corresponds to the interaction technique that uses information from the input devices. Interaction techniques have a tremendous impact on operator performance. Standard interaction techniques encompass complex mechanisms (e.g., modification of the cursor's movement on the screen according to the acceleration of the physical mouse on the desk). This design space is of prime importance and HCI research has explored multiple possibilities for improving performance, such as enlarging the target area for selection on touch screens [16] and providing on-screen widgets to facilitate selection [1].

The interaction mainly takes place though the manipulation of input devices (e.g., keyboard or mouse) and the perception of information from the output devices (e.g., a computer screen or speaker). Another channel usually overlooked is the direct perception by the operator of information produced (usually as a side effect and not on purpose) of the underlying cyber-physical systems (e.g., noise or vibrations from an aircraft engine (represented by the dotted line at the bottom of Fig. 1)).

4 MUMR-MIODMIT: A Technique to Describe the Architecture of Multi User Multi Roles Interactive Systems

As seen in the previous section, H-MIODMIT already allows us to describe complex systems with multiple devices being used by a user thus meeting most of our requirements. However, it is still missing some of the requirements for interactive post-stroke rehabilitation systems:

- It is mono-user
- It takes into account multiple devices, but not multiple systems
- There is no representation of synchronous/asynchronous user interactions.

MUMR (Multi User Multi Role)-MIODMIT is a technique that allows describing all of the remaining points by extending the existing notation. Figure 2 shows the different types of possible connections between architectural parts in both H-MIODMIT and MUMR-MIODMIT. Both can represent dataflows between part and subparts of the system using a plain arrow. However, H-MIODMIT only consider one user, whereas MUMR-MIODMIT consider multiple users. When multiple users interact with a system, their interaction can be completely independent of each other (asynchronous user interaction) or the multiple users may interact together in a synchronous way. MUMR-MIODMIT thus refines the user interaction arrow, by borrowing the synchronous/asynchronous communication of the UML notation [19] and using it for user interaction.

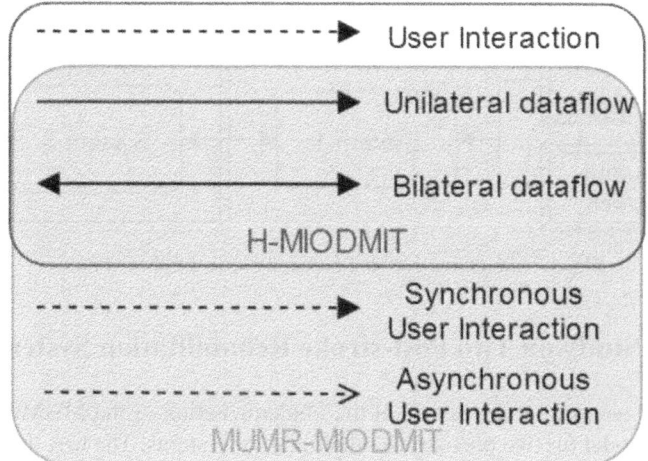

Fig. 2. Synchronous/Asynchronous interactions represented in MUMR-MIODMIT

Figure 3 presents the results of the application of the MUMR-MIODMIT notation to a system that includes several input-output devices and several users. Figure 3 presents three users using the same system, user 1 and 2 using it synchronously with a potential conflict on the I/O device 2 while the user 3, which use the system asynchronously, thus not having any concurrency issues with another user on I/O device 2.

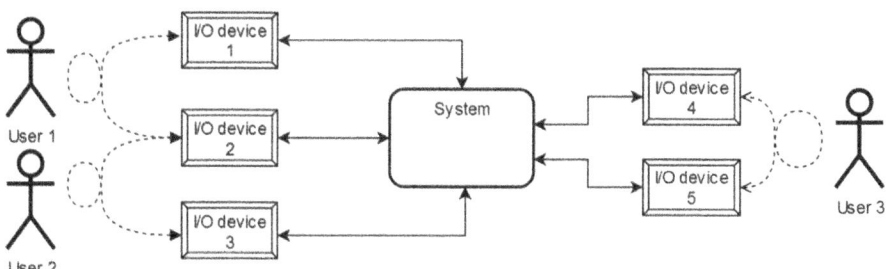

Fig. 3. Multiple users interacting with a system using different input/outputs

Figure 4 presents the element of notation that enables to describe that a system is embedded in a I/O device. This new element of notation enables to represent the multiple systems one or several users may have to interact with.

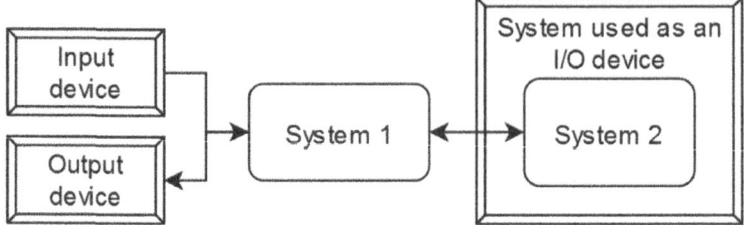

Fig. 4. Interactive system interacting with another system

5 A Case Study on Two Post-stroke Rehabilitation Systems

In this section we present the results of the implementation of the MUMR-MIODMIT architectural model for two post-stroke rehabilitation systems. The first implementation targets a motorized exoskeleton that is usually exploited in physical medicine departments in hospital. Such system requires to be configured and monitored by a trained specialist. The second implementation targets a multi-devices training platform that could be used in different types of environments (hospital, home or medical practices).

5.1 Case Study A: The Armeo-Spring

The first case study presents the results of the application of the MUMR-MIODMIT notation to the Armeo Spring [9]. The Armeo Spring is a passive exoskeleton which provides UE (Upper Extremity) weight support and allows early training of motor skills (presented in Fig. 5). It enables six degrees of freedom and can be attached to the UE at the level of the arm, forearm and wrist. It allows self-initiated arm movements in a large tridimensional workspace and can be configured to provide support against gravity is provided by adjustable springs. The Armeo Spring records kinematic parameters via seven sensors positioned on the different exoskeleton joints, and provides all the exoskeleton joint angles, the effector location in a tridimensional workspace (used to control a cursor on a screen) and the grip pressure.

One of the challenges when designing exoskeleton systems is that you cannot directly use the raw input of the user, this input needs to be calibrated to calculate the correct amount of force, depending on how you want to guide the user, in a limited amount of time, involving multiples interactive systems that interact with each other.

Figure 6 presents the MUMR-MIODMIT model of the Armeo Spring. The user stands on the left side. The user has a cognitive processor, a perceptual processor and a motor processor. All of these processors interact with the components of the long-term and working memory. The perceptual processor stores the perceived Haptic return of the system. The motor processor uses the desired movement that is in the working memory to drive the interaction.

In Fig. 6, the system stands on the right side and is composed of the *Motion capture* part and the *Motion guidance* part. The interaction loop stands between the user and the system and includes an arrow going in the *Sensors* block and an arrow coming from the *Haptic device* block. The *Sensors* block represents the physical hardware of the *Sensor*

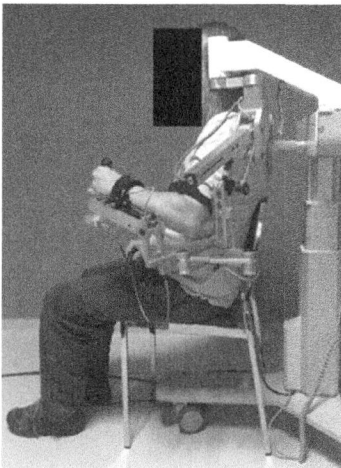

Fig. 5. Installation of a patient performing a training exercise with the Armeo Spring (from [3])

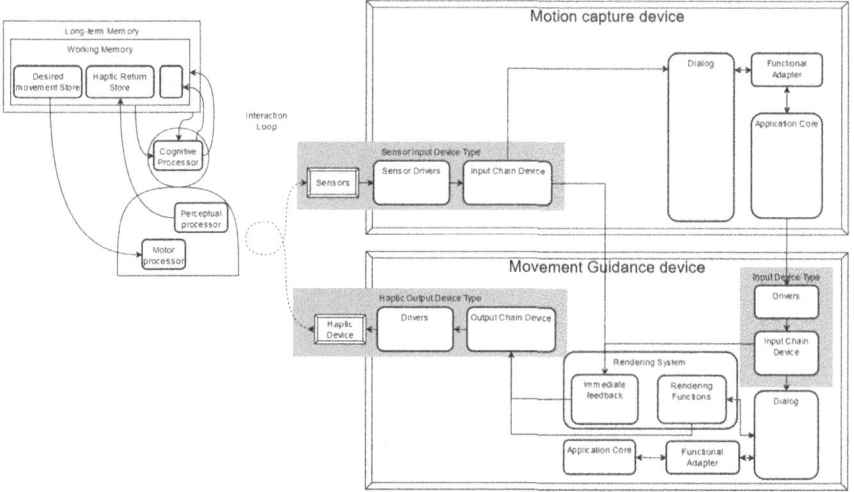

Fig. 6. MUMR-MIODMIT model of a rehabilitation tool chaining multiple I/O devices

input device type of the biggest physical device *Motion Capture*. The *Sensors* block sends a signal to the *Sensors drivers* that sends information to the *input chain device* sends information either in the *rendering system* to the *immediate feedback*, which in turn sends immediately the feedback on to the output devices, or in the *Dialog* part of the system. The dialog communicates with the *Functional Adapter*, which communicates to the *Application Core*. The second part of the system, the *Movement Guidance device* communicates with the *Motion capture device* from two different points: from the *Input Chain Device* to the *Immediate Feedback* module or from the *Application Core* to the *Drivers*.

Inside the motion capture device, the *Dialog* receive data from the *Input Chain Device*, send, and receive data to the *Functional Adapter*. The *Functional Adapter* itself send and receive data to the *Application Core* and the application can send data to the *Drivers* of the *Movement Guidance device*. Inside the *Movement Guidance device*, the *Drivers* send data to the *Input Chain Device*, which send data to either the *Dialog* or the *Immediate Feedback*. The *Dialog* part sends to and receives data from the *Rendering Functions* and the *Functional Adapter*. The *Functional Adapter* sends data to and receive data from the *Application Core*. The *Rendering Functions* and the *Immediate feedback*, which are part of the *Rendering System* module, send data to the *Output Chain Device*. The *Output chain device* sends data to the *Drivers*, which sends data to the physical device *Haptic device*. Finally, the *Haptic Device* is an output in the interaction loop with the user, thus closing the loop.

This case study shows that the two modules act as two different systems interacting with each other, while the user only uses input and output of one system.

5.2 Case Study B: Shape Sorter Drum Task

The second case study presents the results of the application of the MUMR-MIODMIT notation to an electronic and computer-based version of the Shape Sorter Drum Task [22]. The electronic version of the Shape Sorter Drum Task [5] aims to provide guidance to the rehabilitation exercise as well as to monitor patient performance (shown in Fig. 7). On a screen, the patient is asked to put the correct shape in the hole of the same shape. On insertion, a red LED is on if it was incorrect, a green one otherwise. The therapist, in real time, can see on the screen what exercise the patient is doing and its current performance.

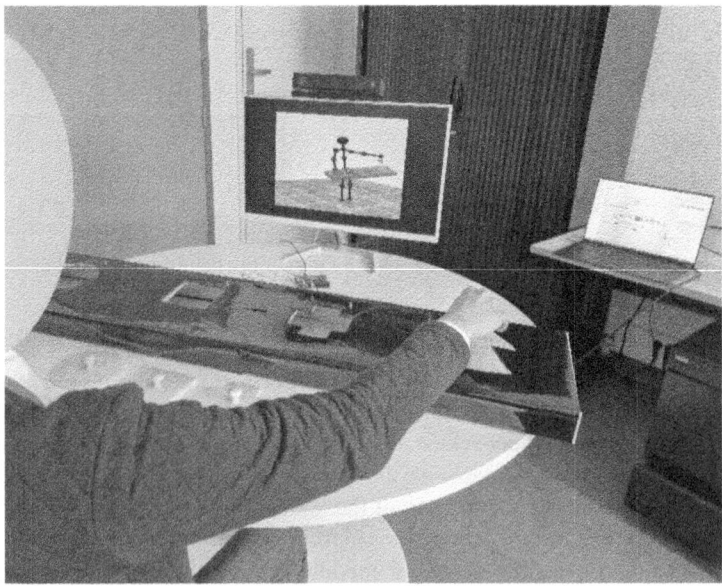

Fig. 7. The electronic and computer-based version of the Shape Sorter Drum Task (from [5])

Figure 8 shows the MUMR-MIODMIT model of the electronic and computer-based version of the Shape Sorter Drum Task, where, this time, multiple users are interacting with it. On the left side the two main users, *Patient* and *Therapist*, are represented. For readability purpose and compared to the representation of the patient in case study A, the representation of the users in case study B are simplified. However, they may be fully detailed like in case study A if needed.

In this case study there are two asynchronous interaction loops (represented by the arrow tip not being full), one for each user, meaning that the two occurs independently of each other.

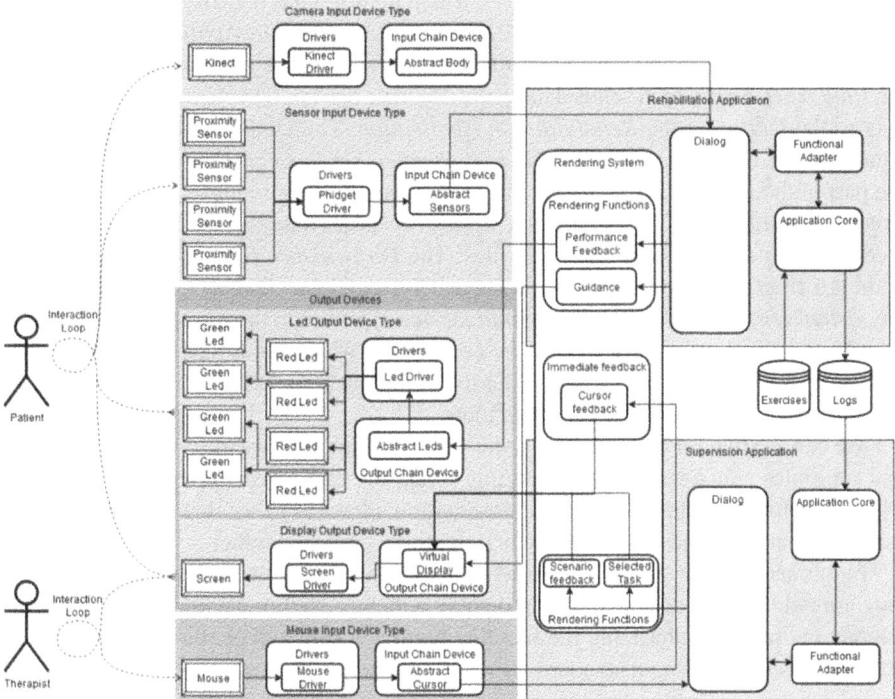

Fig. 8. MUMR model of the Shape Sorter Drum Task

The interaction loop with the therapist uses the *Mouse Input Device type* as input and the *Display Output Device Type* as output, both being used asynchronously, meaning that the interaction does not block the other user interaction. In the *Mouse Input Device Type*, the *Mouse* hardware receives the input and sends data to the *Mouse driver*, which in turns sends data to the *Abstract Cursor* of the *Input Chain Device* module. The Abstract cursor sends data to the *Cursor feedback*, which is part of the *Rendering System* and to the *Dialog* of the *Supervision Application*. Inside the *Supervision Application*, the *Dialog* sends data to and receive data from the *Functional Adapter*. It also sends data to the *Scenario Feedback* and the *Selected tasks*, that are two *Rendering Functions* of the *Rendering System*. The *Functional Adapter* sends data to and receives data from the *Application Core*. The *Application Core* also reads data from the *Logs* part, which is stored data, outside of the application. The *Cursor feedback*, *Scenario feedback* and *Selected Task feedback* of the *Rendering System* are sent to the *Virtual Display* of the *Output Chain Device* of the *Display Output Device Type*. Inside the *Display Output Device Type*, the *Virtual Display* sends data to the *Screen driver* which in turn sends data to the *Screen* hardware, which is an output used by the therapist in the interaction loop.

The other user labelled *Patient* is using two inputs in the interaction loop: *Camera Input Device Type* and *Sensor Input Device Type*. The patient also uses two outputs: *Led Output Device Type* and *Display Output Device Type*. All of them are used in an asynchronous way. In the *Camera Input Device Type*, the *Kinect* hardware receives user input, and sends data to the *Kinect Driver*, which in turn sends data to the *Abstract Body*. The *Abstract Body* sends data to the *Dialog* of the *Rehabilitation Application* part. In the *Sensor Input Device Type*, four *Proximity Sensors* receive the user input and send it to the *Phidget Driver*, which sends data to the *Abstract Sensor*. The *Abstract Sensor* sends data to the *Dialog* of the *Rehabilitation Application*. The Dialog send to and receives data from the *Functional adapter*, the *Performance Feedback*, and the *Guidance* which are part of the Rendering functions. The *Functional Adapter* sends to and receives data from the *Application Core*. The *Application Core* stores data to external *Logs* files and receives data from external *Exercices* files. The *Performance feedback* part sends data to the *Abstract Leds* of the Led Output Device Type and the *Guidance* part sends data to the *Virtual Display* of the *Display Output Device Type* and to the *Abstract Leds* as well. The *Abstract Leds* part sends data to the *Led Driver*, which sends data to four hardware output devices *Green led* and to four hardware output devices *Red Led* which are one of the outputs of the patient interaction loop.

We can see that on this case study, this new technique allows us to represent the fact that multiples sub-systems are interacting along the same systems, how they interact with each other, and the multi-user representation allows us to see that they both use the same input at the same time, which can potentially cause a conflict.

In this case study, we can see that while the two user uses two different applications, they are rendered on the same output display which is used in parallel in two different interaction loops, emphasizing the fact that there is a potential conflict, which was not possible to show if using a standard H-MIODMIT model.

6 Conclusion

MUMR-MIODMIT is a notation and technique to describe the architecture of the interactive systems supporting the training of stroke patients. It is based on the MIODMIT description technique for the architecture of interactive systems and enables to describe the main aspects of a post-stroke rehabilitation-training platform. Such technique supports the analysis and design of software and hardware architectures of post-stroke rehabilitation training platforms. In particular, this technique enables to represent multiple users, multiple systems and asynchronous interactions between them.

Designers and developers of post-stroke training platforms have neither conceptual framework to rely on nor practical APIs (Application Programming Interfaces) to develop HW/SW (HardWare/SoftWare) platforms that exploit several post-stroke rehabilitation technologies. Moreover, existing technologies for post-stroke rehabilitation are not explicitly connected the patients' tasks for the whole rehabilitation program and it is thus difficult to adapt or adjust these technologies according to the recent performance of patients' during previous training session in their rehabilitation program. The development and implementation of several integrated post-stroke rehabilitation technologies within a full and consistent rehabilitation training program requires being able to engineer training platforms with multiple HW and SW components and to engineer coordinated and relevant connections between these components and user tasks. The precise and unambiguous description of training platform architecture supports the analysis of possible conflicts between users on parts of the architecture (e.g., synchronous conflicting access on one device used synchronously). The proposed technique thus aims to inform the design and development of training platforms for post-stroke rehabilitation. Future work will investigate how this technique can be integrated within a user-centered design and development process.

References

1. Albinsson, P.A., Zhai, S.: High precision touch screen interaction. In: Proceedings of the ACM CHI Conference, pp. 105–11 (2003)
2. Bass, L., et al.: A metamodel for the runtime architecture of an interactive system: the UIMS tool developers workshop. SIGCHI Bull. **24**(1), 32–37 (1992)
3. Brihmat, N., Loubinoux, I., Castel-Lacanal, E., Marque, P., Gasq, D.: Kinematic parameters obtained with the ArmeoSpring for upper-limb assessment after stroke: a reliability and learning effect study for guiding parameter use. J. Neuroeng. Rehabil. **17**(1), 130 (2020). https://doi.org/10.1186/s12984-020-00759-2
4. Canny, A., Bouzekri, E., Martinie, C., Palanque, P.: Rationalizing the need of architecture-driven testing of interactive systems. In: Bogdan, C., Kuusinen, K., Lárusdóttir, M., Palanque, P., Winckler, M. (eds.) HCSE 2018. LNCS, vol. 11262, pp. 164–186. Springer, Cham (2019). https://doi.org/10.1007/978-3-030-05909-5_10
5. Carayon A., et al.: Engineering rehabilitation: blending two tool-supported approaches to close the loop from tasks-based rehabilitation to exercises and back again. Proc. ACM Hum.-Comput. Interact. **7**(EICS), 23 (2023). Article 177. https://doi.org/10.1145/3593229
6. Cronel, M., Dumas, B., Palanque, P., Canny, A.: MIODMIT: a generic architecture for dynamic multimodal interactive systems. In: Bogdan, C., Kuusinen, K., Lárusdóttir, M., Palanque, P., Winckler, M. (eds.) HCSE 2018. LNCS, vol. 11262, pp. 109–129. Springer, Cham (2019). https://doi.org/10.1007/978-3-030-05909-5_7
7. Diaper, D., Stanton, N.A.: The Handbook of Task Analysis for Human-Computer Interaction. Lawrence Erlbaum, Mahwah (2004). ISBN 0-8058-4432-5
8. Fitts, P.: The information capacity of the human motor system in controlling the amplitude of movement. J. Exp. Psychol. **47**, 381–391 (1954)
9. Gijbels, D., Lamers, I., Kerkhofs, L., et al.: The Armeo spring as training tool to improve upper limb functionality in multiple sclerosis: a pilot study. J. Neuroeng. Rehabil. (2011).https://doi.org/10.1186/1743-0003-8-5
10. Kousidis, S., Kennington, C., Baumann, T., Buschmeier, H., Stefan, K., Schlangen, D.: A multimodal in-car dialogue system that tracks the driver's attention. In: Proceedings of the 16th International Conference on Multimodal Interaction (ICMI 2014), pp. 26–33. ACM, New York (2014)
11. Kraleva, R., Kralev, V.: On model architecture for a children's speech recognition interactive dialog system. In: Proceedings of International Scientific Conference on Mathematics and Natural Sciences (2009). https://arxiv.org/pdf/1605.07733

12. Kolski, C., Forbrig, P., David, B., Girard, P., Tran, C.H.I.D.U.N.G., Ezzedine, H.: Agent-based architecture for interactive system design: current approaches. Perspect. Eval. **5610**, 624–633 (2009). https://doi.org/10.1007/978-3-642-02574-7_70
13. Langhorne, P., Bernhardt, J., Kwakkel, G.: Stroke rehabilitation. Lancet **377**, 1693–1702 (2011). https://doi.org/10.1016/s0140-6736(11)60325-5
14. Langhorne, P., Coupar, F., Pollock, A.: Motor recovery after stroke: a systematic review. Lancet Neurol. **8**, 741–754 (2009)
15. Murre, J.M.J., Dros, J.: Replication and analysis of Ebbinghaus' forgetting curve. PLoS ONE **10**(7) (2015). https://doi.org/10.1371/journal.pone.0120644
16. Olwal, A., Feiner, S.: Rubbing the Fisheye: Precise Touch-Screen Interaction with Gestures and Fisheye Views. Conference Supplement of UIST 2003, pp. 83–84 (2003)
17. Forbrig, P., Bundea, A., Kühn, M.: Challenges with traditional human-centered design for developing neurorehabilitation software. In: Bernhaupt, R., Ardito, C., Sauer, S. (eds.) HCSE 2022. LNCS, vol. 13482, pp. 44–56. Springer, Cham (2022). https://doi.org/10.1007/978-3-031-14785-2_3
18. Forbrig, P., Bundea, A., Pedersen, A., Platz, T.: Digitalization of training tasks and specification of the behaviour of a social humanoid robot as coach. In: Bernhaupt, R., Ardito, C., Sauer, S. (eds.) HCSE 2020. LNCS, vol. 12481, pp. 45–57. Springer, Cham (2020). https://doi.org/10.1007/978-3-030-64266-2_3
19. Rumbaugh, J., Booch, G., Jacobson, I.: The Unified Modeling Language Reference Manual: Covers UML 2.0. Addison-Wesley, Upper Saddle River (2006)
20. Soukoreff, W., MacKenzie, S.: Towards a standard for pointing device evaluation, perspectives on 27 years of Fitts' law research in HCI. IJHCS. **61**(6), 751–789 (2004)
21. The top 10 causes of death. In: World Health Organization. https://www.who.int/news-room/fact-sheets/detail/the-top-10-causes-of-death. Accessed 26 May 2023
22. Weiss, T., et al.: Deafferentation of the affected arm: a method to improve rehabilitation? Stroke. **42**(5), 1363–1370 (2011). https://pubmed.ncbi.nlm.nih.gov/21454817/
23. World Stroke Organization. 2022. Global Stroke Fact Sheet 2022. https://www.world-stroke.org/assets/downloads/WSO_Global_Stroke_Fact_Sheet.pdf. Accessed Feb 2023

Serious Game for Company Governance: Supporting Integration, Prevention of Professional Disintegration and Job Retention of People with Disabilities

Yosra Mourali[1], Benoit Barathon[2], Maxime Bourgois Colin[3], Sondes Chaabane[1], Raja Fassi[3], Alice Ferrai[4], Yohan Guerrier[1], Dorothée Guilain[4], Christophe Kolski[1(✉)], Yoann Lebrun[5], Sophie Lepreux[1], Philippe Pudlo[1], and Jason Sauvé[6]

[1] Univ. Polytechnique Hauts-de-France, CNRS, UMR 8201 – LAMIH, 59313 Valenciennes, France
{Yosra.Mourali,christophe.kolski}@uphf.fr
[2] Emploi et Handicap Grand Lille, 23 chemin du Moulin Delmar, 59708 Marcq en Baroeul Cedex, France
[3] HandiexpeRh, 40, rue Eugène Jacquet, 59708 Marcq en Baroeul, France
[4] Handyn'Action, 12 Boulevard Froissart, 59300 Valenciennes, France
[5] Serre Numérique, 2 rue Péclet, Parc des Rives Créatives, BP 80577, 59 308 Valenciennes Cedex, France
[6] ARACT Hauts-de-France, 197, Rue Nationale, 59000 Lille, France

Abstract. Despite efforts made, it is still difficult to clearly grasp the situation of disability within companies. Therefore, to overcome this problem, this paper proposes and describes an innovative tool that may facilitate corporate governance. The tool is a serious game developed on an interactive RFID tabletop associated with tangible objects. The serious game is called SG-HANDI. It is designed to be used in disability awareness sessions in companies. It implies directors, managers and employees and proposes a set of challenges to be solved as a team. The objective is to encourage stakeholders to think and act collectively for a more accessible company.

Keywords: Disability · Company Governance · Accessibility · Serious Game · Interactive Tabletop · Tangible Interaction

1 Introduction

In an increasingly diverse world, inclusive governance is essential to enable all people to fully participate in society. As people with disabilities represent a significant portion of the global population, it is important for companies to reflect this diversity. In this regard, two inclusion policies are adopted by countries to require companies to implement accessibility measures and provide equitable employment opportunities for people with disabilities. The first one is based on laws and legislation, as is the case in France and Italy. This inclusion policy imposes a quota of disabled employees on companies. This obligation is often accompanied by subsidies or interesting tax deductions. However, the second policy is based on the promotion of disabled people by relying on the responsibility of business leaders, as is the case in UK and USA.

Thus, by engaging people with disabilities, today's companies aspire to respect legislation and diversity, which can improve their reputation with consumers and investors. In this context, this paper proposes an innovative IT tool (serious game) aiming to encourage companies integrate people with disabilities and help them to manage the adequate working environment. Precisely, the serious game is developed on an RFID interactive tabletop called *TangiSense* (built by the Rfidees company) associated with tangible objects. It is inspired by previous work (see for instance [1–3]). It is used to carry out awareness-raising actions for company stakeholders (executives, managers and employees able-bodied and with disabilities) [4, 5].

The serious game is a part of the SG-HANDI project "(Serious Game (SG) de sensibilisation à l'intégration à la prévention de la désinsertion professionnelle et au maintien dans l'emploi des personnes en situation de HANDIcap)" translated by (Serious Game (SG) to raise awareness of integration, prevention of professional disinsertion and retention in employment of people with HANDIcap). The project is led by ARACT Hauts-De-France in partnership with the LAMIH-UMR CNRS 8201 laboratory, the Serre Numérique (CCI Grand Hainaut), the Handyn'Action association, the Emploi et Handicap Grand Lille association and the HandiexpeRH consulting firm. This project is supported by Agefiph.

The rest of the paper is structured as follows: Sect. 2 presents key concepts about employment, disability and governance. Section 3 exposes the serious game conceptual modeling. Section 4 and Sect. 5 are devoted to the implementation and the evaluation of the SG-HANDI serious game. Finally, this paper is ended by a conclusion and perspectives.

2 Background

Generally, questions about the integration and/or retention in employment of people with disabilities or health difficulties arise when the person is being integrated or when needs to be retained. Regarding this matter, governance strategies can vary enormously from one company to another and from one country to another. They often require global solutions, both technical (hardware, software, etc.) and organizational. Recommendations, case studies, and best practices are available, see for example [6–9].

For companies needing to innovate or transform in this field, a raising awareness phase among the company's stakeholders is essential in order to better reflect on global solutions. These solutions can lead to anticipatory (technical and/or organizational), individual and collective actions. This paper focuses on the phase of disability awareness in the work environment.

Sensitivity to disability in the workplace can be facilitated through various activities. Several websites offer recommendations on this topic [10–12]. For example, Thriiver suggests eight disability awareness activities, including (1) organizing an inclusion day, (2) offering employees the opportunity to share their experiences, (3) providing key resources, (4) creating an accessibility map, (5) running workshops and training days, (6) organizing charity fundraising events, (7) offering staff workplace needs assessments, and (8) providing mental health awareness training sessions. In this paper, we focus on activity number five: running workshops and training days.

It is possible to organize workshops inside or outside the company, for raising awareness of disability in the workplace. These workshops are often based on classical pedagogical methods (conferences, presentation of case studies, exercises, and so on). What about incorporating new interaction techniques to keep participants engaged and promote active learning?

As far as we are concerned, our approach is based on a serious game [13, 14]. Serious games provide a novel approach that combines technology, storytelling, and gameplay mechanisms [15]. They also can provide interactive visuals and audio to create a dynamic learning environment accommodating different preferences. Thus, serious games can enhance the overall effectiveness of the workshop and facilitate deeper understanding and engagement with the subject matter. The serious game proposed in this paper is used in a collective context. This context involves (1) one or several animators, specialized in management of disability in companies, and (2) participants from the company. The approach proposed is described below.

3 Proposal

This section presents users and then the flow of SG-HANDI serious game.

3.1 The SG-HANDI Serious Game Users

The SG-HANDI serious game is designed to be used in corporate disability awareness sessions. The target users have two profiles:

(1) The facilitator specialized in employment and disability. The facilitator is the master and manages the serious game, gives instructions, orchestrates and moderates discussions to make players aware in the target domain.
(2) The players are company stakeholders such as executives, managers and employees able-bodied and with disabilities. They play as a team to solve challenges proposed by the serious game. They are the people to raise awareness.

3.2 SG-HANDI Serious Game Flow

In order to facilitate the description of the SG-HANDI serious game flow, the activity diagram is shown in Fig. 1.

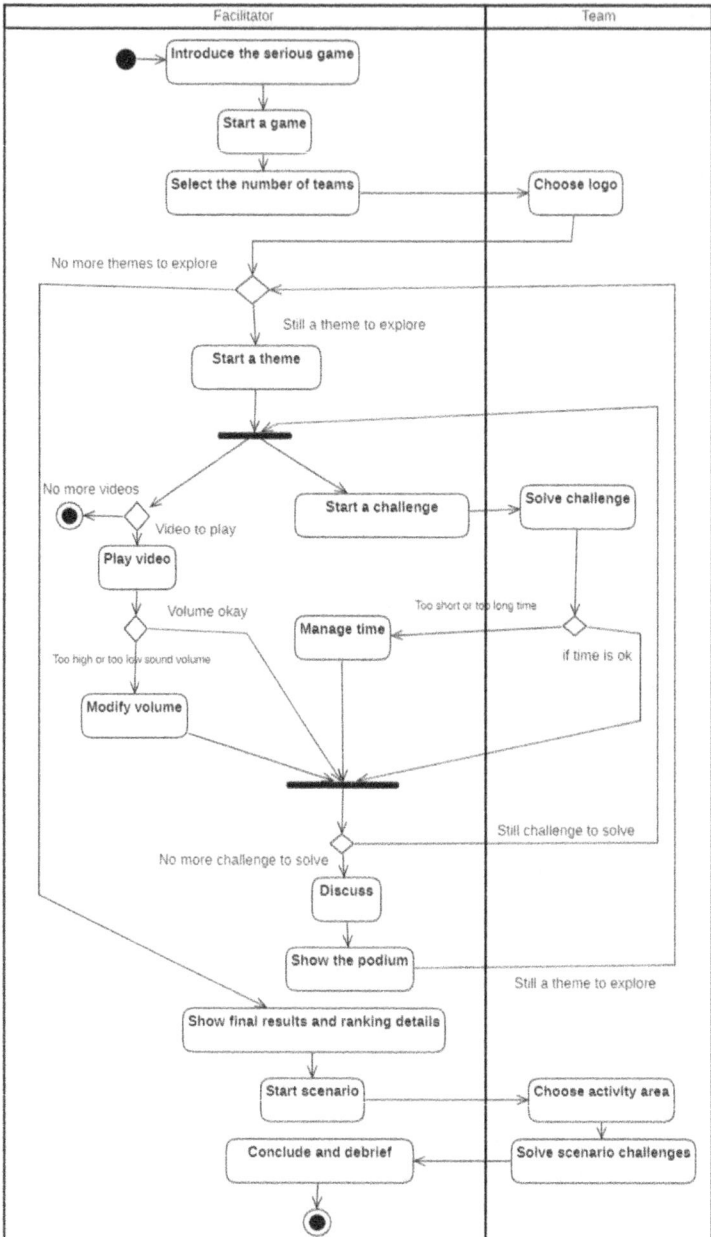

Fig. 1. Activity diagram of the SG-HANDI serious game.

At the beginning of an awareness session, the facilitator presents the game. The facilitator explains the technical aspects and shows how to use the interactive tabletop and the tangible objects. Then, the facilitator launches the game. Once the players are ready, the facilitator decides to start the game by placing a personal pawn on the tabletop. The facilitator asks the participants (players) to get into two or three teams depending on their number. Each team chooses its (tangible) pawn and its (digital) logo. At this time, the facilitator starts the first part of the game including five themes. Each specific tangible object is equipped with an RFID tag.

In each theme, players are faced with one or more challenges (for instance the 6th challenge of the 1st theme is "In your opinion, who are the people with disabilities, Antoine the masseur with a sight problem, Mireille the dyslexic manager or Bob the handyman with osteoarthritis of fingers?") to solve in time. Each challenge is most often followed by complementary explanations and/or a discussion. The facilitator can add or reduce the time granted for the resolution of the challenges by using two different tangible objects. For some themes, video sketches are planned. With the pawn, the facilitator can take breaks and replay. The facilitator also can increase or decrease the sound volume thanks to a remote control. During the serious game, the facilitator decides to move from one challenge to another and from one theme to another using the pawn. Generally, a theme is closed by the podium and a discussion which recapitulates and enriches what has just been seen. Once all themes are covered, the facilitator shows final results and ranking details. At this moment the second part of the game can begin.

The first part lasts about 1 h 30. If there is enough time to continue the session, in the second part of the game, participants choose together a sector to be approached through a real-life scenario: logistics, industry or commerce. Here, players put themselves in the shoes of a company that aims to recruit a new employee. To accomplish this mission, six challenges must be met.

At the end (whether the session consisted of one or two parts), the facilitator proposes a debriefing and concludes the awareness-raising session.

4 Implementation

4.1 The SG-HANDI Serious Game Design

As explained in Sect. 3, the SG-HANDI serious game consists of two parts. Before starting, the facilitator asks players to form a certain number of teams (two or three), each with its pawn and its logo. In the first part, teams confront each other with challenges covering several facets of the disability field such as basic concepts, legislation, company resources, professional disinsertion, compensation, etc. Challenges are grouped into five themes and each theme is closed by a podium: (1) situational disability, (2) disability typologies, (3) disability recognition, (4) compensation and (5) job displacement. The second part of the SG-HANDI serious game concerns a real-life scenario. Participants try to succeed in the hiring mission of a new employee. First of all, they have to choose one of these sectors: logistics, commerce, or industry. Then, they have to accomplish five challenges: (1) job description, (2) job offer, (3) the interview, (4) communication and (5) testimony.

The themes and messages were chosen in working groups with experts in the field. Themes and messages are inspired by the experience and recurring questions encountered by SG-HANDI project partners (working to promote the inclusion of vulnerable people, particularly people with disabilities) during their interventions. The various challenges were selected on the basis of playful activities that the SG-HANDI project partners had previously implemented and which had proved successful.

4.2 The SG-HANDI Serious Game Characteristics

In our design approach, we have turned to diversify (1) tangible objects, (2) types of questions, as well as (3) the temporality in order to avoid monotony.

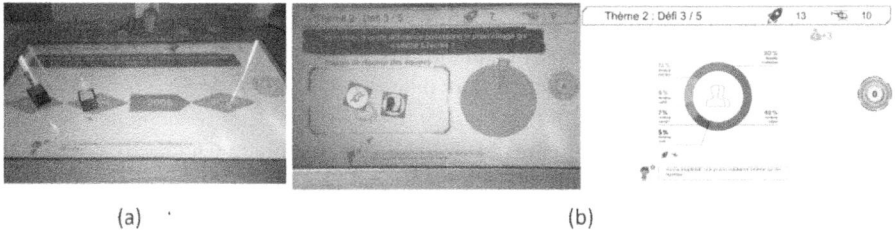

(a)   (b)

Fig. 2. (a) Players use pawns to solve the 3rd challenge of the 2nd theme (quiz) and result is provided directly after time runs out, (b) players use disability logos to solve the 3rd challenge of the 3rd theme (drog and drop) and results are provided at the end. *"Thème 1: Défi 3/6 - Selon vous, en quelle année est sortie la dernière loi en vigueur sur le handicap?"* translated by: "Theme 1: Challenge 3/6 - When do you think the latest disability law?", "Voici le récapitulatif, vous pouvez maintenant debriefer sur vos réponses!" translated by: "Here is the summary, you can debrief your answers now!", *"Thème 2: Défi 3/5 - Selon vous, à quelle déficience correspond le pourcentage qui s'affiche à l'écran?"* translated by: "Theme 2: Challenge 3/5 - Which disability do you think the percentage shown on the screen corresponds to?", *"Espace de réponse"* translated by: "Response space", "Utilisez les étiquettes de la boite LOGOS pour les poser sur la table. Vous ne pouvez poser qu'un seul logo à la fois." translated by: "Use the labels in the LOGOS box to place them on the table. You can only place one logo at a time.", *"5% Handicap visuel, 7% Handicap mental, 8% Handicap auditif, 13% Handicap psychique, 20% Maladies invalidantes, 45% Handicap moteur"* translated by: "5% Visual disability, 7% Mental disability, 8% Hearing disability, 13% Psychological disability, 20% Disabling diseases, 45% Motor disability", "Voici le récapitulatif, vous pouvez maintenant debriefer sur vos réponses!" translated by: "Here is the summary, you can debrief your answers now!"

Thus, players can use tangible objects printed on paper such as terms, disabilities logos, poker chips as well as tangible objects in the form of a 3D-printed characters (pawns). These different types of objects are used to solve different kinds of questions (challenges) such as quiz and drog and drop. From challenge to challenge the temporality is varied to avoid dullness and boredom. For example, in the third challenge of the first theme (Fig. 2(a)), as soon as the time is up, the score is assigned to teams. However, in the third challenge of the second theme (Fig. 2(b)), participants must choose a type of handicap based on a percentage, without knowing the other percentages, and only receive the result at the end of all the questions. Moreover, a countdown with background sound has been integrated into the game to motivate players.

5 Evaluation

First series of the serious game uses started at the end of March 2023. Three groups of participants were involved during disability awareness workshops, with the same animator. The number of participants was respectively 10 (Group 1), 8 (Group 2) and 10 (Group 3). The duration of each workshop was approximately one and a half hours (with a focus on the first part of the serious game, due to a temporal constraint).

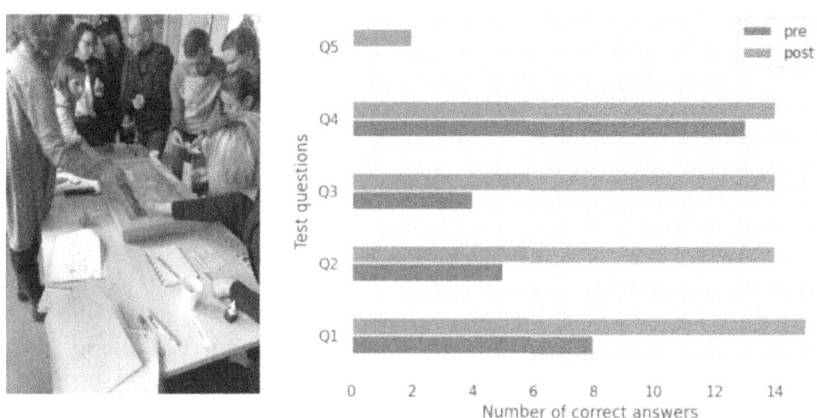

Fig. 3. (a) A situation giving insight into participant engagement. The facilitator is visible on the right side. (b) Pre-post test results.

For evaluation purposes, the sessions were filmed. A consent form as well as authorization for image rights were signed by each participant (15 participants) before the workshop. Furthermore, pre-post test (consisting of 5 questions) and satisfaction questionnaires were addressed to each of them. The pre-test and the post-test were used to evaluate and compare the knowledge of the participants before and after the serious game use. To do this, the questions were formulated in the same way for both tests: (Q1) Which of these proposals never existed?, (Q2) In your opinion, what is the proportion of invisible disabilities?, (Q3) In your opinion, when was the latest disability law?, (Q4) What proportion of disabilities do you think occur in adulthood and childhood? and (Q5) In your opinion, how many out of 8 disabled people have a job?

For the satisfaction questionnaire, SUS (System Usability Scale) was used [16, 17]. The facilitator also completed an evaluation questionnaire after the three workshops: CSUQ (Computer Usability Satisfaction Questionnaires) [16, 18].

The first observations showed a promising engagement of the participants (see Fig. 3(a)). Comparison of pre-test and post-test results by question shows a significant progression in post-test (as seen in Fig. 3(b)) proving the serious game effectiveness. Several possible improvements were also proposed by the facilitator: for instance it would be useful to shortcut one or several collective challenges, in case the available time for a workshop is not sufficient. Other part of collected data is being analyzed.

6 Conclusion and Future Work

The management of disability in companies leads to new important challenges for corporate governance. To prepare different types of action, it is important to raise the awareness of various stakeholders (e.g., executives, managers, staff representatives, employees) to integration, prevention of professional disintegration and job retention of people with disabilities. One of the possible activities is to organize workshops for raising awareness of disability in the workplace. For supporting such workshops, a serious game on RFID interactive tabletop with tangible objects was proposed. A first prototype has been realized.

The first uses of the serious game started at the end of March in several companies. Three groups of participants were involved in disability awareness workshops.

These first uses led to promising results, showing a commitment of the participants throughout the game. Several improvements of the serious game were also highlighted. The perspectives concern first of all the analysis of all the data collected during these first evaluations. It would also be useful to study how activities related to serious games can be articulated with other awareness-raising activities suggested in [11].

As part of our upcoming study, we plan to conduct interviews with facilitators who have gained experience in both conventional facilitation methods and the use of serious games on interactive tabletop. The aim of these interviews is to gain insight into the impact of technology, particularly serious games, during awareness sessions, and compare the two approaches.

Acknowledgment. The authors are very grateful to Agefiph for the financial support provided to the SG-HANDI project. They thank also the facilitators and participants who were present at the awareness sessions.

References

1. Lebrun, Y., Vispi, N., Lepreux, S., Chaabane, S., Kolski, C.: Simulation on an RFID interactive tabletop with tangible objects of future working conditions: prospects for implementation in the hospital sector. In: Chaabane, S., Cousein, E., Wieser, P. (eds.) Healthcare Systems: Challenges and Opportunities, pp. 115–127. ISTE & Wiley, London (2022)
2. Vispi, N., Lebrun, Y., Lepreux, S., Chaabane, S., Kolski, C.: Simulation on RFID interactive tabletop of working conditions in industry 4.0. In: Borangiu, T., Trentesaux, D., Leitão, P., Cardin, O., Lamouri, S. (eds.) SOHOMA 2020. Studies in Computational Intelligence, vol. 952, pp. 343–354. Springer, Cham (2021). https://doi.org/10.1007/978-3-030-69373-2_24
3. Lebrun, Y., Adam, A., Mandiau, R., Kolski, C.: A model for managing interactions between tangible and virtual agents on an RFID interactive tabletop: case study in traffic simulation. J. Comput. Syst. Sci. **81**(3), 585–598 (2015)
4. Mourali, Y., et al.: Conception centrée utilisateur d'un jeu sérieux sur table interactive avec objets tangibles: application au handicap en entreprise. 34ème Conférence Internationale Francophone sur l'Interaction Humain-Machine (2023)
5. Mourali, Y., et al.: Design and prototyping of a serious game on interactive tabletop with tangible objects for disability awareness in companies. AAATE: Assistive Technology: Shaping a sustainable and inclusive world (2023)

6. Harder, H.D., Geisen, T.: Disability Management and Workplace Integration International Research Findings. Routledge, London (2016)
7. ILO: Business as unusual: Making workplaces inclusive of people with disabilities. Working Paper No 3. Bureau for Employers' Activities and Skills and Employability Department, Geneva (2014). https://www.ilo.org/wcmsp5/groups/public/---ed_emp/---ifp_skills/documents/publication/wcms_316815.pdf
8. ILO: Disability in the Workplace, Company Practices. Working Paper No 3, Bureau for Employers' Activities and Skills and Employability Department, Geneva (2010). https://www.ilo.org/wcmsp5/groups/public/@ed_emp/@ifp_skills/documents/publication/wcms_150658.pdf
9. Riley, C.: Disability and Business: Best Practices and Strategies for Inclusion. UPNE, Lebanon (2006)
10. ERI: 10 Ways for Employers to Promote a Disability-Friendly Workplace. Advocacy, Featured, News (2020). https://eri-wi.org/blog/2020/10/01/national-disability-awareness-month-2020/
11. Wolberger, L.: Ways to Promote Disability Awareness in the Workplace. Forbes (2022). https://www.forbes.com/sites/forbestechcouncil/2022/09/13/ways-to-promote-disability-awareness-in-the-workplace/?sh=7f3616463f24
12. Thriiver: Your Guide to Disability Awareness in the Workplace (2023). https://thriiver.co.uk/disability-awareness-in-the-workplace/
13. Alvarez, J., Djaouti, D.: Serious Game: An Introduction. Questions Théoriques (2012)
14. Alvarez, J., et al.: A formal approach to distinguish games, toys, serious games & toys, serious re-purposing & modding and simulators. IEEE Trans. Games **15**, 399–410 (2022)
15. Dimitriadou, A., Djafarova, N., Turetken, O., Verkuyl, M., Ferworn, A.: Challenges in serious game design and development: educators' experiences. Simul. Gaming **52**(2), 132–152 (2021). https://doi.org/10.1177/1046878120944197
16. Lewis, J.R.: Measuring perceived usability: the CSUQ, SUS, and UMUX. Int. J. Hum.-Comput. Interact. **34**(12), 1148–1156 (2018). https://doi.org/10.1080/10447318.2017.1418805
17. Brooke, J.: SUS: a "quick and dirty" usability scale. In: Jordan, P.W., Weerdmeester, B., Thomas, A., Mclelland, I.L. (eds.) Usability Evaluation in Industry, pp. 189–194. CRC Press (1996)
18. Lewis, J.R.: IBM computer usability satisfaction questionnaires: psychometric evaluation and instructions for use. Int. J. Hum.-Comput. Interact. **7**(1), 57–78 (1995). https://doi.org/10.1080/10447319509526110

Two Concepts of Domain-Specific Languages for Therapists to Control a Humanoid Robot

Peter Forbrig(✉), Alexandru Umlauft, Mathias Kühn, and Anke Dittmar

Chair of Software Engineering, University of Rostock, Albert-Einstein-Str. 22, 18055 Rostock, Germany

{peter.forbrig,alexandru-nicolae.umlauft,mathias.kuehn, anke.dittmar}@uni-rostock.de

Abstract. Domain-Specific Languages allow domain experts to specify their knowledge in such a way that software can be generated based on those specifications. In the context of a project using a humanoid robot Pepper as coach in therapies of post stroke patients we developed the language TaskDSL4Pepper and StateDSL4Pepper. They allow therapists to specify the behavior of a robot based on the two concepts of task models and hierarchical state machines.

The paper discusses both languages that support commands like `say<text>`, `show<image>`, `play<video>` and `raiseArms`. First feedback from therapists in comparing both languages is discussed.

Keywords: Domain-Specific Languages · Humanoid Robot Pepper · Behavior

1 Introduction

The goal of our project E-BRAiN [7, 9] (Evidence-Based Robotic Assistance in Neurorehabilitation) was the development of software for a humanoid robot Pepper to support patients after a stroke. The first therapy session has always to be performed by a human therapist. They give an introduction and analyze several attributes of patients.

We use a software architecture with MQTT messages [18]. Mishra et al. [16] discuss applications of this technology, which allows secure communication via a server that allows the registration of clients to specific topics.

The programing language Python was used to implement several therapies. In parallel, we worked on domain-specific languages to allow therapists to manipulate specifications of exercises. The examples of a mirror therapy will be used to show its specification in both languages. A mirror therapy is useful for patients that cannot control their arm. Exercises are performed with the not affected arm. Patients have to look into a mirror and have to imagine that the handicapped arm is moving. This trains the brain in such a way that the handicapped arm can be used again after a while. Figure 1 gives an impression of the therapy situation.

The paper is structured in such a way that we first have a look on related work that considers programming approaches of end-users for humanoid robots. Afterwards we introduce our domain-specific languages. A discussion of comparing both approaches follows. Finally, there is a paragraph that summarizes the paper and provides an outlook.

M. Harrison et al. (Eds.): EICS 2023 Workshops, LNCS 14517, pp. 50–59, 2024.
https://doi.org/10.1007/978-3-031-59235-5_6

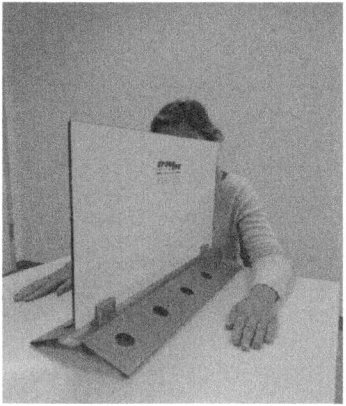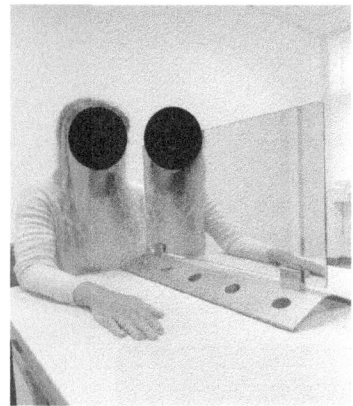

Fig. 1. Therapy situation for the mirror training.

2 Related Work

A survey on end-user robot programming is presented by Ajaykumar et al. in [1]. They discuss 121 references. However, most of the references are related to general robots. Only two papers have the term Humanoid in their title. One of them is Moros et al. [17] that uses a visual block concept with Scratch. However, there is no big difference to programming like in Python. The other paper is Leonardi et al. [14]. The authors suggest a trigger-action-programming for end-user developers. This approach is also discussed in [15]. It allows users to manipulate rules that control the behavior of robots.

Coronado suggests with different co-authors in [3] and [4] a block structured programming approach for end-user. They discuss Node Primitives (NEP), a robot programming framework aimed at enabling the creation of usable, flexible and cross-platform end-user programming (EUP) interfaces for robots and a system that is called Open RIZE. Figure 2 gives an impression how the programming environment of OPEN RIZE looks like.

The DSL CommonLang (Rutle et al. [21]) is created for writing code for different robots. It is a kind of abstract programming language that is designed for programmers and not for end users.

The languages Kinematics-DSL, Motion-DSL, and Transforms-DSL are introduced in Frigerio et al. [11]. They are used in connection with the code generator RobCoGen. These languages are focused on motions and are very technical oriented. They are also not designed for end users.

The same is true for a DSL discussed in Heinzemann and Lange [12]. They discuss a DSL for specifying robot tasks that is intended to support programmers.

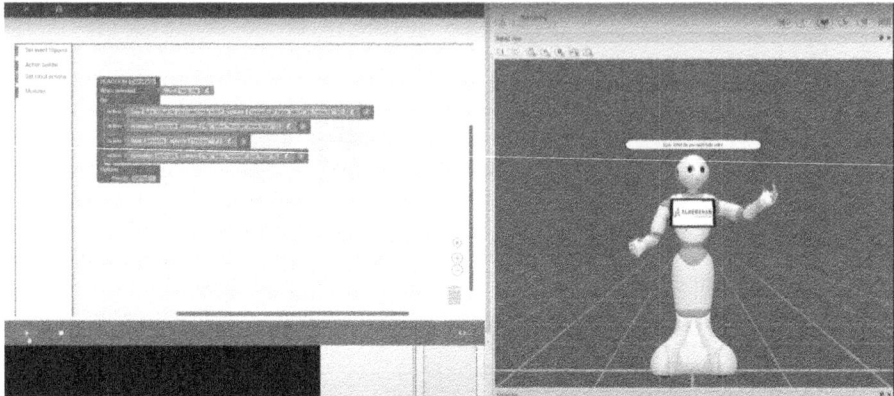

Fig. 2. The RIZE end-user programming interface (on the left) and the robot simulator (on the right) from [22].

Oishi et al. [20] discuss end-user programming for humanoid robots. However, not much details are provided.

A DSL for integrating neuronal networks into robot control is presented by Hinkel et al. [13]. Transfer functions can be specified in a domain-specific way in Python and through an XML format. We used the concept of providing specific predefined functions in Python for an interpreter of state machines in our implementations as well. However, end users cannot update those specifications.

The markup language MPML-HR (Multi-modal Presentation Markup Language for Humanoid Robots,) is presented by Nishimura et al. [19]. It allows end-users to specify the interaction between humanoid robots and humans. However, it is still a markup language. Nevertheless, such a language might be helpful as well.

3 The Domain Specific Language TaskDSL4Pepper and StateDSL4Pepper

3.1 TaskDSL4Pepper

For task models there exists already the textual domain-specific language CoTaL [8]. It is implemented using the tool Xtext [23] that generates syntax driven textual editors.

Results of this editor can be used to generate code using the tool Xtend [24]. Together with other options code generation can be performed for the environment CoTaSE [5] that allows the animation of model instances. CoTaSE is discussed in detail in Buchholz et al. [2].

TaskDSL4Pepper is an extension of this language by extending it with specific tasks of a robot like say <text>, show <picture>, play <video>, raiseArms, etc. Additionally, there exists an environment that allows the animation of model instances and sends MQTT messages to Pepper if such a robot command is animated.

```
team coop {
    root training = greeting >> train{*} [> end_exercises
              task greeting = pepper.greet |=| patient.greet
}
role patient {
  root train = greet >> listens >>
                perfoms_exercises{*} [> finishes_exercises >> bye
    task perfoms_exercises = perform_correct [] perform_wrong
}
robot pepper {
  root armtraining= greet >> training{*} [> end_exercises
     task training = introduce >> train
       task introduce = say introduction ||| show startPict
       task train = play exerVid ||| say three >>
                 wait 10 >> train_imagine
         task train_imagine = say ten ||| imagine
           task imagine = wait 15 >> say look >> say imagine
      task end_exercises = say bye
}
text greeting     = "Good morning.";
text introduction = "Let us start to perform our second exercise"
image startPict   = "st_pic_10_1.jpg"
video exerVid     = "st_10_1.mp4"            // *    Iteration
text imagine   = "Imagine it is your arm"    // >>   Enabling
text look      = "Look at mirror"            // [>   Disabling
text three     = "Train three times"         // |||  Interleaving
text ten       = "Train ten times"           // []   Choice
text bye       = "Bye, till next time"
```

Fig. 3. Specification of a mirror therapy in TaskDSL4Pepper.

The specification of Fig. 3 consists of four parts. First, there is the team model that is reduced to a minimum here. Second, there is a role model of a patient, which specifies that the training starts with the task greet. It is followed by listens and an iteration of the task perform_exercises. The iteration is stopped by executing the task finishes_exercises. A refinement of perform_exercises specifies that it can be performed correctly or wrong. Third, the tasks of a robot pepper are specified. The training with the arm starts with greet. This is followed by an iteration of training that is stopped by executing end_exercises. Each iteration starts with an introduction followed by the training.

The introduction consists of saying an introduction and presenting a picture showing in parallel the start and end position of the exercise. The further training starts with showing a video of the exercise. In parallel the robot says: "Train three times", waits ten seconds and performs the task train_imagine. The robot waits fifteen seconds, says: "look at mirror" and "Imagine it is your arm". At the end of all exercises the robot says: "Bye, till next time".

As fourth part there exists the declaration of texts, pictures and videos. A list of available pictures and videos on the robot can be used to select them in the editor.

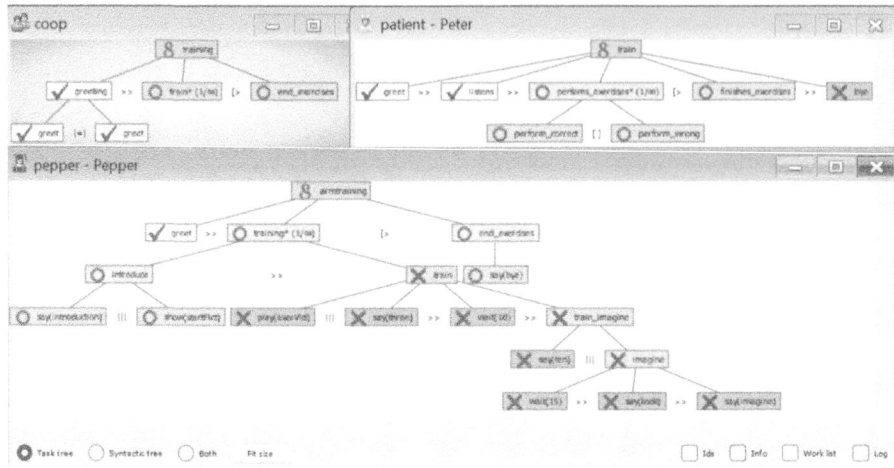

Fig. 4. Animated instances of task model after the greeting phase.

One can see in Fig. 4 at the top left the animated team model that reflects the fact that the greeting task was already performed. At the top right there exists the animated model of patient Peter that already greeted and listened. Peter can perform next an exercise correctly or wrong. At the bottom there exists the animated model of robot Pepper. The robot greeted already and can say the instruction and show a picture in parallel. This can be seen at the very left button of the animated task model. According to the specification the training can also be ended by clicking on the say(bye)-button of the task model of pepper.

The task structure seems to be better visible in the graphical representation than in the text form. However, this view can be generated automatically for different tools.

4 StateDSL4Pepper

Our second domain-specific language is based on the concept of hierarchical state machines. It allows currently only commands of a robot as actions inside a state. In this way, hierarchical Moore automata are supported. Besides normal states there exist initial and final states. A final state does not have any actions.

Transitions can be executed immediately or delayed after a certain timespan. They also can be triggered by named events that do not have to be declared beforehand.

Each specification in StateDSL4Pepper results in a code generation to the programming language Java. During execution of those programs MQTT messages can be sent to a robot. For events a window pops up that provides the different alternatives of the current state transition. A user has to select one of those events.

Figure 5 specifies the similar therapy as Fig. 3 and provides an impression of the state-based DSL.

```
initial greet{
    say greeting
    ->introduce
}
state introduce{
    say introduction
    show startPict
    ->train
}
state train{
    play exerVid
    say three
    ->train_imagine after 10.0 s
}
state train_imagine{
    say ten
    ->imagine after 15.0 s
}
state imagine{
    say look
    say imagineThat
    ->end_exercise
}
state end_exercise{
    say bye
    event new->greet
    event end->ende
}
final ende                  // -> State transition
```

Fig. 5. Specification of a mirror therapy in StateDSL4Pepper.

The declarations of texts, pictures and videos are the same as for task models and were therefore omitted here. One can see in the specification two delayed transitions that are executed after 10 and 15 s. Additionally, there exists one state with two event transitions that result in the following window (Fig. 6).

Fig. 6. User interface for triggering events.

56   P. Forbrig et al.

Our software architecture allows the integration of different robots. Only an interpreter for MQTT messages is necessary. Therefore, we implemented a very simple virtual robot that allows us to evaluate the results of specifications without having access of a real robot. Figure 7 presents the situation of the virtual robot after `show startPict` before the transition to the state `train` is performed.

Fig. 7. Situation of the virtual robot after `show startPict`.

4.1 Comparison of Both Languages

To compare both languages, we performed one evaluation session in the clinical environment of the University of Greifswald with two participants that were not familiar with programming concepts. However, they knew the mirror therapy. We started with an introduction of the concepts of task models and state charts that lasted about one hour. Afterwards, the here presented examples were discussed. However, they were in German language.

The first question was about the task greeting in Fig. 4. What has to be changed that the robot really says something. The participants were able to do that. The same is true for introducing a task that has to be inserted after the greeting activities for the robot. Pepper should say: "If we continue in the same way as in the previous therapy session, we will nearly reach your private goal till Christmas". Both participants were able to perform this task for both kinds of specifications. Some further similar tasks were discussed about one hour.

Finally, a questionnaire was presented to the participants. They had to fill a form with four comparison questions. There was a final fifth question that allowed to provide any hints. The first four questions were provided in form of a table for both languages. It was expected that participants enter values between zero and ten. However, this worked for the first two questions only after some comments and did not work for the two last questions. Participants provided a star only.

1. How do you like the language concept?
2. How do rate the readability of both examples?

3. Do you feel confident to update a specification of a therapy by yourself?
4. Do you feel confident to create a new specification of a new therapy exercise?

The concept of states was scored higher than that of tasks (tasks 8 from 10, states 10 from 10). The readability was scored "good" for both approaches (10 from 10). Also for question three and four participants were sure to be able to fulfill the task. However, we were not able to really check this.

It is not easy to find therapists that volunteer for such evaluations. We felt that we cannot treat them like students and asked for a meeting from 10:00 to 12:30 only.

There also was the opportunity to express some ideas for improvements. Participants had some problems with the temporal operators of the task models. More time to become familiar with those operators would be necessary. Participants asked for a list of keywords with their semantics. Keywords like root, task, state and event (they were not translated) seem to be not as well-known as for computer scientists.

Besides the limitations of the low number of therapists there exists the constraint of the limited "training time". Maybe, more time for discussing the concepts is necessary. Additionally, we demonstrated the task animator and the running Java programs. This might have influenced the evaluation as well. The running Java programs are much simpler than the animated task models. It might take some time to recognize advantages of these detailed animation. We are afraid that we did not evaluate the languages but the running systems. It might be worth to encapsulate the task model instances by a simple user interface with only activatable tasks like that for states. Some more experiments seem to be necessary. However, finding therapists that are willing to spend time for such experiments is challenging.

5 Summary and Outlook

The paper discussed the idea of end-user programming for humanoid robots by two different textual domain-specific languages TaskDSL4Pepper and StateDSL4Pepper. As the names already express one language is following the concept of task models while the other one specifies hierarchical Moore state machines. As most references of the literature we used the term end-user programming. However, the end users are the patients and not the therapists. We have a nested design space like Dittmar and Hensch [6] discussed for personas.

A first feedback from therapists provided an advantage of StateDSL4Pepper. However, there were discussed several limitations for that. Nevertheless, the approach of textual domain-specific languages seems to be promising, even that we did not find similar approaches in the literature.

It is our intention to further extend both languages. Generic components are under development for both languages. The same is true for variables and functions for StateDSL4Pepper. There is also a need for a further temporal relation that allows to ask a patient whether he is ready. If he does not react the question is repeated again. This is also done a third time until an error state is reached that provides an alarm. Additionally, sensor data can be included into the models. First ideas are discussed in Forbrig et al. [10].

Acknowledgements. This joint research project "E-BRAiN - Evidence-based Robot Assistance in Neurorehabilitation" is supported by the European Social Fund (ESF), reference: ESF/14-BM-A55-0001/19-A01, and the Ministry of Education, Science and Culture of Mecklenburg-Vorpommern, Germany. The sponsors had no role in the decision to publish or any content of the publication.

References

1. Ajaykumar, G., Steele, M., Huang, C.-M.: A survey on end-user robot programming. ACM Comput. Surv. **54**(8), 36 (2021). Article 164. https://doi.org/10.1145/3466819
2. Buchholz, G., Forbrig, P.: Extended features of task models for specifying cooperative activities. Proc. ACM Hum.-Comput. Interact. **1**(EICS), 1–21 (2017). Article No. 7. https://doi.org/10.1145/3095809
3. Coronado, E., Mastrogiovanni, F., Venture, G.: Design of a human-centered robot framework for end-user programming and applications. In: Arakelian, V., Wenger, P. (eds.) ROMANSY 22 – Robot Design, Dynamics and Control. CISM International Centre for Mechanical Sciences, vol. 584, pp. 450–457. Springer, Cham (2019). https://doi.org/10.1007/978-3-319-78963-7_56
4. Coronado, E., et al.: Towards a modular and distributed end-user development framework for human-robot interaction. IEEE Access **9**, 12675–12692 (2021)
5. CoTaSE. https://www.cotase.de/. Accessed 30 Mar 2022
6. Dittmar, A., Hensch, M.: Two-level personas for nested design spaces. In: Proceedings of the 33rd Annual ACM Conference on Human Factors in Computing Systems (CHI 2015). Association for Computing Machinery, New York, pp. 3265–3274 (2015). https://doi.org/10.1145/2702123.2702168
7. E-BRAiN Homepage. https://www.ebrain-science.de/en/home/. Accessed 29 Mar 2022
8. Forbrig, P., Dittmar, A., Kühn, M.: A textual domain specific language for task models: generating code for CoTaL, CTTE, and HAMSTERS. In: EICS 2018 Conferences, Paris, France, pp. 5:1–5:6 (2018)
9. Forbrig, P., Bundea, A., Bader, S.: Engineering the interaction of a humanoid robot pepper with post-stroke patients during training tasks. In: EICS 2021, pp. 38–43 (2021)
10. Forbrig, P., Umlauft, A., Kühn, M.: A Domain-Specific Language for Prototyping the Behavior of a Humanoid Robot that Allows the Inclusion of Sensor Data, submitted to HCII 2023, Copenhagen, Denmark (2023)
11. Frigerio, M., Buchli, J., Caldwell, D.G., Semini, C.: RobCoGen: a code generator for efficient kinematics and dynamics of articulated robots, based on domain specific languages. J. Softw. Eng. Robot. **7**(1), 36–54 (2016)
12. Heinzemann, C., Lange, R.: vTSL - a formally verifiable DSL for specifying robot tasks. In: 2018 IEEE/RSJ International Conference on Intelligent Robots and Systems (IROS), pp. 8308–8314 (2018)
13. Hinkel, G., Groenda, H., Vannucci, L., Denninger, O., Cauli, N., Ulbrich, S.: A domain-specific language (DSL) for integrating neuronal networks in robot control. In Proceedings of the 2015 Joint MORSE/VAO Workshop on Model-Driven Robot Software Engineering and View-Based Software-Engineering (MORSE/VAO 2015), pp. 9–15. Association for Computing Machinery, New York (2015). https://doi.org/10.1145/2802059.2802060
14. Leonardi, N., Manca, M., Paternò, F., Santoro, C.: Trigger-action programming for personalising humanoid robot behaviour. In: Proceedings of the 2019 CHI Conference on Human Factors in Computing Systems (CHI 2019), pp. 1–13. Association for Computing Machinery, New York (2019). Paper 445. https://doi.org/10.1145/3290605.3300675

15. Manca, M., Paternò, F., Santoro, C.: Analyzing trigger-action programming for personalization of robot behaviour in IoT environments. In: Malizia, A., Valtolina, S., Morch, A., Serrano, A., Stratton, A. (eds.) IS-EUD 2019. LNCS, vol. 11553, pp. 100–114. Springer, Cham (2019). https://doi.org/10.1007/978-3-030-24781-2_7
16. Mishra, B., Kertesz, A.: The use of MQTT in M2M and IoT systems: a survey. IEEE Access **8**, 201071–201086 (2020). https://doi.org/10.1109/ACCESS.2020.3035849
17. Moros, S., Wood, L., Robins, B., Dautenhahn, K., Castro-González, Á.: Programming a humanoid robot with the scratch language. In: Merdan, M., Lepuschitz, W., Koppensteiner, G., Balogh, R., Obdržálek, D. (eds.) RiE 2019. Advances in Intelligent Systems and Computing, vol. 1023, pp. 222–233. Springer, Cham. https://doi.org/10.1007/978-3-030-26945-6_20
18. MQTT. https://mqtt.org/. Accessed 04 Dec 2022
19. Nishimura, Y., et al.: A markup language for describing interactive humanoid robot presentations. In: Proceedings of the 12th International Conference on Intelligent User Interfaces (IUI 2007), pp. 333–336. Association for Computing Machinery, New York (2007). https://doi.org/10.1145/1216295.1216360
20. Oishi, Y., Kanda, T., Kanbara, M., Satake, S., Hagita. N.: Toward end-user programming for robots in stores. In Proceedings of the Companion of the 2017 ACM/IEEE International Conference on Human-Robot Interaction (HRI 2017), pp. 233–234. Association for Computing Machinery, New York (2017). https://doi.org/10.1145/3029798.3038340
21. Rutle, A., Backer, J., Foldøy, K., Bye, R.T.: CommonLang: a DSL for defining robot tasks. In: Proceedings MoDELS 2018 Workshop MORSE (2018). http://star.informatik.rwth-aachen.de/Publications/CEUR-WS/Vol-2245/morse_paper_1.pdf
22. Tian, L., et al.: Redesigning human-robot interaction in response to robot failures: a participatory design methodology. In: Extended Abstracts of the 2021 CHI Conference on Human Factors in Computing Systems (CHI EA 2021), pp. 1–8. Association for Computing Machinery, New York (2021). Article 57. https://doi.org/10.1145/3411763.3443440
23. Xext. https://www.eclipse.org/Xtext/. Accessed 30 Nov 2022
24. Xtend. https://www.eclipse.org/xtend/documentation/index.html. Accessed 02 Dec 2022

Engineering Interactive Systems Embedding AI Technologies (EIS-embedding-AI Workshop)

An Approach to Leverage Artificial Intelligence for Car-Parking Related Mobile Applications

Alba Bisante[✉], Venkata Srikanth Varma Datla, Gabriella Trasciatti, Stefano Zeppieri, and Emanuele Panizzi

Department of Computer Science, Sapienza University of Rome, Rome, Italy
bisante@di.uniroma1.it

Abstract. This paper describes an approach that combines smartphones, context awareness, implicit interaction, and machine learning to ease the design, distribution, and updating of AI-powered features. By following a series of steps, including identifying user needs, determining the scope of AI intervention, collecting and training data, incorporating user feedback, and continuously updating the model that runs in the smartphone, we design context-aware car parking apps that provide personalized and practical assistance to drivers. The proposed approach is particularly relevant when minimizing distractions and optimizing user support is crucial.

Keywords: Artificial Intelligence · Context-Awareness · Implicit Interaction · Machine Learning · Mobile Applications · Task Facilitation · User Feedback

1 Introduction

Mobile applications have become an integral part of our daily lives [14], and the potential for leveraging artificial intelligence (AI) to facilitate and enhance user experiences within these applications is vast. AI can support or replace repetitive and demanding tasks, offering users a more varied and efficient interaction with the application [2]. However, despite the increasing attention and interest in AI across various fields, integrating AI-based functionality into mobile applications poses unique design and implementation challenges, such as individuating the correct set of features that can benefit from AI and the gathering of quality data for training. Moreover, the limited processing power and memory must be considered, as well as battery life.

Our research mainly focuses on context-aware smartphone apps designed for the smart parking context. This domain presents a compelling opportunity to harness the power of AI to provide valuable features and services to users, who will no longer need to enter data by hand but can rely on AI to understand their needs. Many drivers rely on their smartphones to assist them in performing tasks while on the road, but this can pose significant safety risks as actively using the smartphone distracts them from driving. We aim to mitigate distractions and enhance the driving experience by leveraging AI-based features in context-aware parking applications. Such apps can provide relevant and timely suggestions by intelligently adapting to the user's situation, making parking-related tasks a seamless experience.

Our group also used this approach in other contexts, such as more critical situations, such as analyzing the smartphone microphone to understand whether a person around it is breathing in post-earthquake scenarios [6].

This paper presents our approach to addressing the challenges of integrating AI into mobile applications. We strongly focus on smartphones, which are highly versatile and powerful enough to collect and process contextual data. Once a user need is identified, we use context data to train dedicated machine learning models. The models, running directly on the smartphone, help predict the user's needs, facilitate the execution of tasks, and sometimes directly replace the user. Examples of parking-related features we addressed by adopting the described approach are identifying events such as parking, un-parking, and approaching a parked car [8,10]; classifying parking types (parallel, diagonal or perpendicular parking; private vs public parking) [4,13]; detecting cruising for parking [7]; estimating the level of parking availability on a street-level scale [3], etc.

2 Engineering

This work presents the development of an interactive system designed for the domain of parking, which we also experimented with in the context of earthquakes [5]. In order to minimize distractions, our smart system can intelligently adapt to the user's situation, automatically understanding the context (driving, looking for parking, parked, etc.) and providing parking recommendations. In our system, we prioritize the user's involvement, creating a collaborative environment that provides a positive user experience and continuous system improvement. Making an AI-powered application where the user feels in control is not an easy task; in the following sections, we report our experience on how users respond to different methods of feedback requests from the system and what we found to be both engaging for the user and feasible for the system.

Moreover, we present a systematic approach to creating AI smartphone applications that describe the entire process, from identifying automatable features to data gathering, model learning, and release.

3 Approach

Our approach can be summarized in four steps:

1. identify the task that can be automated and the degree of AI involvement. In order to select the right task, the users' behavior and needs must be thoroughly analyzed to understand where automation is functional. On the other

hand, AI involvement depends on the type of available data and the capabilities of devices.
2. Data collection for training. We release a companion application that collects users' feedback on the task, creating the first dataset used to train the machine learning model.
3. Train a machine learning model over gathered data. This passage can be iterated when a significant new amount of data is available and if the application has a personalized approach and thus must adapt to users' habits.
4. Release the app and allow user feedback to refine the data labeling. This allows for continuous updates to the running model.

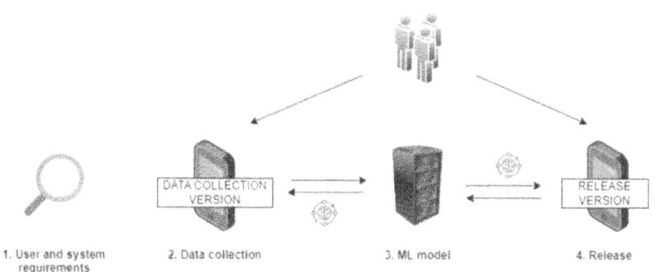

Fig. 1. A visualization of our learning loop

3.1 Identifying the Task

Identifying tasks that can be automated while ensuring a positive user experience can be challenging [11]. It is crucial to strike a balance where users feel in control of the system while benefiting from AI-powered automation.

During the identification phase, it is essential to begin by understanding the user's needs. This step involves gaining insights into their habits, needs, challenges, and areas where automation could enhance their experience. By empathizing with the user and comprehending their requirements, we can lay the foundation for effective AI integration.

Subsequently, it is vital to determine to which extent AI can intervene by identifying the appropriate model and the necessary features and data. This step entails evaluating the task's complexity and the feasibility of using AI.

For example, a significant user need is getting help to find a parking spot. We thoroughly analyzed how drivers look for on-street parking places and identified average behaviors in cruising for parking [7]. AI can help detect when a driver is looking for parking and prompt the system to look up a database for a suitable parking spot, anticipating an explicit request, hence a distraction on the user's part and possibly saving cruising time. We designed an interaction that keeps the user in control by not forcing them to accept parking spot suggestions and letting them ignore any requests on the AI part (e.g., a notification advising the user that a nearby parking facility has available spots).

Identifying the Context. Our approach to developing context-aware mobile apps for the driving context emphasizes assisting users as naturally and seamlessly as possible. Leveraging the power of smartphones, we capitalize on their capability to collect a diverse range of contextual data and computing power for model processing. Important data that can be gathered from smartphones include readings from inertial sensors (such as the accelerometer, gyroscope, magnetometer, and GPS), weather conditions, temperature, time zone, lighting conditions, noise, and specific user-related information like visited places, calendar events, and nearby users.

In order to tailor our services specifically to the driving context, we have developed an implicit interaction approach [8,10] to identify when the user enters and exits a vehicle. Our solution leverages the smartphone's sensing capability of users' locations and motion activities and merges them to infer parking and un-parking events, avoiding any explicit interaction on the user's part. More in detail, such an approach relies on the machine learning models integrated into the smartphone, enabling recognition of the user's activity type (driving or walking) and simple decision algorithms to refine the models' detection, further enhancing their accuracy and augmenting the activity detected with a location and a timestamp. By isolating the driving context, we can provide personalized and relevant services accordingly, triggering even more context-specific features. For example, the above-cited cruising detection feature is triggered only if the user is detected to be driving. Currently, we are developing another AI that focuses on recognizing the user's means of transport, such as cars, buses, trams, or metro. Such an AI can augment and refine the detection capabilities currently available on standard smartphones, based on the device bluetooth connections.

3.2 Gather Data

Training machine learning models necessitates a substantial amount of data. Consequently, our standard practice is to develop a companion version of the target application to facilitate the collection and labeling of data.

By distributing a companion app, we can streamline the data collection process. Such apps incorporate HCI principles [12], ensuring ease of understanding, intuitive interactions, and a visually appealing interface. Adhering to these guidelines enhances user engagement, encourages user participation, and ultimately contributes to the quality of the collected dataset. Indeed, when users find the application easy to understand and navigate, they are more likely to provide accurate and meaningful labels, enhancing the overall effectiveness of the training dataset.

For the means of transport recognition project, the data collection app we developed warns users when it detects that they have entered a state of motion on a motorized vehicle. Upon receiving the notification, users can respond promptly by selecting the specific type of transport used, enabling the collection of labeled training data. However, developing this app presented its challenges. One of the critical difficulties we encountered was that users occasionally overlooked the notification and needed to input the data manually later. To mitigate this, we introduced an editable past travel history feature in the data-collection app that allows users to label the collected data after the trip is finished (see Fig. 2). This enhancement lets users label their data retrospectively, ensuring a more complete and accurate dataset.

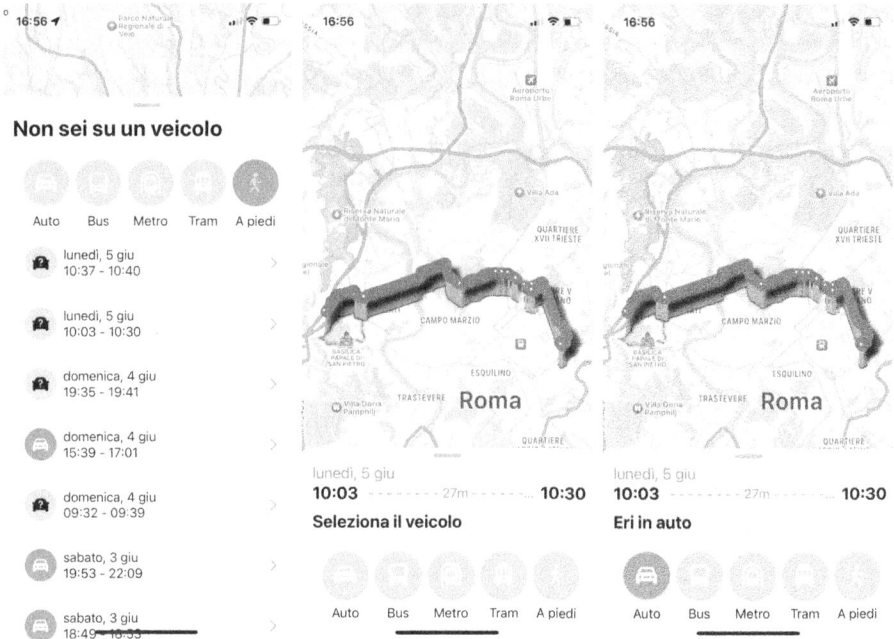

Fig. 2. Example screens from the data collection app to recognize the user's means of transport. In this version, the app does not yet implement a recognition model. Therefore, the labeling of each trip is done manually by the user. On the left: "You are not in a vehicle". The user can assign a vehicle type to each of their previous trips. Center: "Select the vehicle". Upon selecting a trip from the history, the trace of collected GPS locations is shown to the user, along with the day and time it took place, as a reminder. Right: "You were in a car". The user labeled the collected data by selecting the means of transport.

3.3 Model Training and Distribution

We train our AI models on dedicated servers. We train a prediction model on the data collected by the companion app; once the model reaches the target accuracy, it is distributed to the apps and can be run locally. As described in the next section, the released application will continue to gather data that can be used to update the model over time.

In the future, we plan to implement personalized applications that learn the habits of individual users; in this case, each user will have a different model,

Fig. 3. Example screen from the released parking app. On the left: "Type of parking". The app predicted the type of parking "angle". On the right: the user can correct the prediction by selecting a different type of parking, "perpendicular", "angle", or "parallel".

which will be refined over time using only the feedback given by that user. Our approach is to design and develop models suitable to be run directly on the user's device. This approach minimizes reliance on external computing resources and prioritizes user privacy in the final application. Indeed, data privacy can be upheld by keeping the data stored on the user's device and only occasionally transmitting it to the server for future model updates. Moreover, it removes the delay that would introduce the communication towards an external resource, allowing for better use of models designed for critical contexts.

3.4 Release, Feedback, and Human-AI Collaboration

Once the model is distributed and operational, we prioritize the user's involvement and empowerment by allowing them to offer feedback on the generated predictions. This feedback mechanism is crucial for collecting valuable insights and identifying potential issues or inaccuracies in the model's output. By actively encouraging user feedback, we establish a collaborative environment where users play an active role in improving the system's performance and enhancing their experience.

Furthermore, we acknowledge the significance of transparency and user control in AI-driven systems. To ensure users always remain in charge of the system and clearly understand its operations, we enable users to revert any changes made by the AI easily. Through such a collaborative approach, we address potential issues, improve the accuracy of future predictions, and cultivate a trustworthy and user-friendly design.

Regarding this matter, we collected valuable feedback directly from users during the test phases of an app that gathered different parking-related features. Users expressed their appreciation when the underlying AI understood context and user actions. For example, they were pleased when the AI accurately recognized the location and time of parking events and the specific type of parking. In contrast, users did not respond favorably when prompted to input information manually. For example, our app initially asked for the type of parking each time a parking event was detected (the aim was to involve the user regardless of the model's guess). However, we learned that this approach needed to be better received. We removed the notification and allowed the model to make an independent guess. We then made the model classification visible to users, enabling them to correct it if necessary. Users positively received this modification, as it gave them more autonomy and control over the data input process without the burden of repetitive requests for interaction. This process of keeping users in charge of the system and making them understand what is happening by also giving the option to undo actions taken by the AI is a crucial step to allow them to recover, understand, and avoid errors [9].

4 Updating Models

The model can be updated by re-training it over newly collected data using user feedback. This is useful in two cases: when the app is released, and thus the availability of data is limited, and when the app must keep track of each user's habits. In the first case, the model is re-trained over the previous data plus the new feedback, and this is repeated over time until the model's precision stops growing. On the other hand, if the application must keep track of individual habits, the model will be re-trained daily in a sliding window fashion, where older data is gradually no longer used to make predictions. In any case, the updated model is distributed to the apps, allowing them always to give the most recent and accurate predictions. To do so, we need to have in place a system to pre-process the collected data, run the model training, save it, and let the apps know that there is a new version of the model available and that they need to update their local version.

We implement the described system as follows:

1. Run a Python script to download the collected data and pre-process it to organize it for the training phase.
2. Train the new model using the new pre-processed data.
3. Save the newly trained model and notify the server that a new version is available.
4. Update the latest model version number on the server to ensure that consequent requests to check for the model version by the apps will trigger the model update task [1].
5. (Optional) Send a silent notification to the app to let it know a new model version is available. This can trigger the model update task directly without the user waiting for it when the app gets opened next time.

This process enables us to enhance the model's accuracy over time and ensure that the data classification aligns with the desired outcomes.

5 Conclusions

This paper proposes an engineering approach to integrate AI into mobile applications, specifically within the context of car parking. Our research demonstrates the potential of AI in enhancing user experiences and facilitating tasks within mobile applications. Exploiting this approach, which leverages context awareness, implicit interaction, and machine learning, we have designed an AI-powered application that provides personalized and practical assistance to drivers.

Our approach addresses the unique challenges of AI integration into mobile applications, such as identifying tasks that can benefit from AI, balancing user experience with AI automation, and gathering quality data for training while preserving the user's privacy. We have shown that with careful design and continuous model updating, it is possible to create an app that meets evolving user needs.

In the context of smart parking, we illustrate the potential that AI applications can provide by giving real-time advice, enhancing the driving experience, and mitigating safety risks associated with active smartphone use while driving.

The described approach is generalizable to different domains that can benefit the application of Artificial Intelligence to mobile User Interfaces to obtain implicit and context-aware interaction. We successfully applied it to another context, post-earthquake presence recognition, demonstrating its versatility and potential for broader application.

References

1. Apple: downloading and compiling a model on the user's device. https://developer.apple.com/documentation/coreml/downloading_and_compiling_a_model_on_the_user_s_device
2. Autor, D.H.: Why are there still so many jobs? the history and future of workplace automation. J. Econ. Perspect. **29**(3), 3–30 (2015)
3. Bassetti, E., Berti, A., Bisante, A., Magnante, A., Panizzi, E.: Exploiting user behavior to predict parking availability through machine learning. Smart Cities **5**(4), 1243–1266 (2022)
4. Bassetti, E., Luciani, A., Panizzi, E.: ML classification of car parking with implicit interaction on the driver's smartphone. In: Ardito, C., et al. (eds.) INTERACT 2021. LNCS, vol. 12934, pp. 291–299. Springer, Cham (2021). https://doi.org/10.1007/978-3-030-85613-7_21
5. Bassetti, E., Panizzi, E.: Earthquake detection at the edge: Iot crowdsensing network. Information **13**, 195 (2022). https://doi.org/10.3390/info13040195
6. Bassetti, E., Panizzi, E., Cavallaccio, G., De Marsico, M.: Human presence detection after earthquakes: an AI-based implicit user interface on the smartphone (2023)
7. Bisante, A., , Panizzi, E., Zeppieri, S.: Cruising-for-parking detection on the smartphone based on implicit interaction and machine learning (2023)
8. Bisante, A., Datla, V.S.V., Zeppieri, S., Panizzi, E.: Implicit interaction approach for car-related tasks on smartphone applications - a demo. In: Proceedings of the 2022 International Conference on Advanced Visual Interfaces. AVI 2022, Association for Computing Machinery, New York, NY, USA (2022). https://doi.org/10.1145/3531073.3534465, https://doi.org/10.1145/3531073.3534465
9. Bisante, A., Dix, A., Panizzi, E., Zeppieri, S.: To err is AI (2023)
10. Bisante, A., Panizzi, E., Zeppieri, S.: Implicit interaction approach for car-related tasks on smartphone applications. In: Proceedings of the 2022 International Conference on Advanced Visual Interfaces. AVI 2022, Association for Computing Machinery, New York, NY, USA (2022). https://doi.org/10.1145/3531073.3531173
11. Martinie, C., Palanque, P.: Task models based gameful design as a mean to increase engagement with automation. In: Workshop on Engaging with Automation (AutomationXP 2022) co-located CHI 2022, No. Session 1: Engaging with Automation: Common Challenges in 3154, CEUR-WS. org (2022)
12. Moreno, A.M., Seffah, A., Capilla, R., Sanchez-Segura, M.I.: HCI practices for building usable software. Computer **46**(04), 100–102 (2013)
13. Panizzi, E., Bisante, A.: Private or public parking type classifier on the driver's smartphone. In: 12th International Conference on Emerging Ubiquitous Systems and Pervasive Networks/11th International Conference on Current and Future Trends of Information and Communication Technologies in Healthcare, vol. 198, pp. 231–236 (2022). https://doi.org/10.1016/j.procs.2021.12.233, https://www.sciencedirect.com/science/article/pii/S1877050921024728
14. Uğur, N.G., Turan, A.H.: Mobile applications acceptance: a theoretical model proposal and empirical test. Int. J. E-Adopt. (IJEA) **11**(2), 13–30 (2019)

Engineering AI-Similar Designs: Should I Engineer My Interactive System with AI Technologies?

David Navarre[1], Philippe Palanque[1,2(✉)], and Célia Martinie[1]

[1] ICS-IRIT, Université Toulouse III – Paul Sabatier, Toulouse, France
{david.navarre,philippe.palanque,celia.martinie}@irit.fr
[2] FlexTech Chair, ESTIA & Centrale SupElec, Bidart, France

Abstract. This position paper questions the need for systematically embedding AI technologies in interactive computing systems when other programming-based alternatives are available. Indeed, while AI technologies bring critical benefits in some contexts such as patterns recognition or computer vision, they also bring issues that prevent them from more generic use. For instance, their blackbox functioning calls for specific research on its explainability which, instead of solving the problems try to address issues brought both to developers (how does my Machine Learning system works) and users (why do I get this result). In addition, we demonstrate that these technologies are not mature enough yet to address reliability and dependability concerns in safety critical domains where probability of failures must be demonstrated to be in the order of magnitude of 10-E9. The paper thus argues for exploiting validated methods in the area of interactive systems engineering to build interactive systems even though their exhibited behavior seems similar to the ones of AI-engineered systems.

Keywords: Interactive Systems Engineering · AI technologies · Reliability · Dependability · Usability

1 Introduction

Current hype for AI technologies makes people believe that computing technologies are able to perform any information processing task and that every computing system embeds AI technologies as soon as data are processed, information generated or deductions made. This is far from being true but this propaganda from media and AI researchers impacts the way users and developers perceive computing systems.

First, as argued in [1] the underlying activity of computing systems embedding AI technologies maybe be more related to human input than computing-based detection and deduction. Second, it might be unwise and counter productive to exploit AI technologies when not required [2].

Indeed, while AI technologies bring critical benefits in some contexts such as patterns recognition or computer vision, they also bring issues that prevent them from more generic use. For instance, their black box functioning calls for specific research on its explainability which, instead of solving the problems try to address issues brought both to developers (how does my Machine Learning system works) and users (why do I get this result). In addition, we demonstrate that these technologies are not mature enough yet to address reliability and dependability concerns in safety critical domains where probability of failures must be demonstrated to be in the order of magnitude of 10-E9.

For several years now, the European Aviation Safety Agency started to address the issues related to the embedding of AI technologies in aviation with a focus on Machine Learning ones, due to the push from industry (see Fig. 1). Indeed, the promises of AI technologies play a kind of smoke and mirrors game, claiming to solve pressing issues such as shortage of skilled developers by declaring proudly the "end of programming" [24] or ensuring faster development time by automating coding tasks [25], even though some disillusions are already made vocal [26].

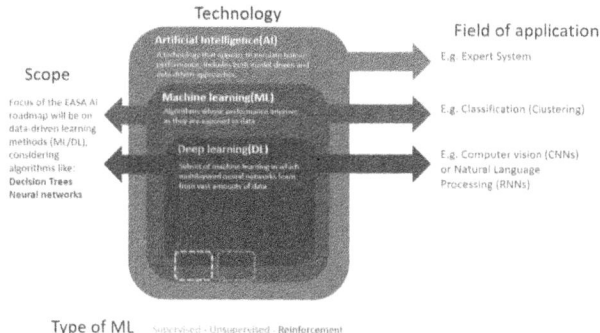

Fig. 1. Types of Machine Learning technologies as presented in [20]

This paper builds on these disillusions and proposes to distinguish between AI-similar designs for interactive systems and engineering interactive systems using AI technologies. We are that, for some designs, exhibiting AI-similar designs does not require the use of AI technologies. Beyond, Sect. 2 highlights the critical issue of explainability of AI technologies while Sect. 3 demonstrates that certification authorities in the civil aviation domain are not going to grant deployment of these technologies due to their intrinsic lack of dependability. To demonstrate the fact that AI-similar designs can be developed using state of the art technologies in interactive systems engineering, we present (in Sect. 4) two simple case studies which might be perceived by developers and users as requiring AI technologies to function. On the first example (fortunettes [3]) we show that Petri-net based behavioral models [4] are able to process state spaces and to forecast possible future states of a user interface. We show in details how the prediction model is constructed based on the original interactive system model [6]. On the second examples (a recommender system for Flight Warning System in a civil aviation commercial aircraft [7]) we show that the recommender system can be described using generic software architectures [8] and that AI technologies are not needed and even forbidden according to European Aviation Safety Agency regulations described in its AI roadmap [5]. Finally, we demonstrate that such AI-similar design may be embedded in critical interactive systems provided that certifiable means (such as formal methods) are used.

2 The Specific and Critical Problem of Explainability in AI-Based Technologies

Explainability refers to the capacity to provide explanations of the decisions of AI algorithms in some level of details [13]. According to them two key aspects have to be considered for an explanation: interpretability and completeness. Interpretability aims to produce descriptions that are understandable for a user, by "using a vocabulary that is meaningful to the user" [13]. Completeness aims to describe the behavior of the AI algorithm in an accurate way and this description includes information to allow "the behavior of the system to be anticipated in more situations" [13].

Several explainability methods co-exist and Speith [14] recently proposed a unified taxonomy for these methods according to the following categories: stage, scope, output format, functioning, result and others. Figure 2 presents a diagram that provides an overview of the Speith taxonomy of explainability methods. The stage can be ante or post hoc explainability. Ante-hoc means that the AI model is directly explainable, and post-hoc means that additional AI models are created to produce explanations of the first AI model (which is, in this case, usually called a black box model). Scope of the explanation can be local or global. Local means that an explanation concerns a single prediction, and global means that an explanation concerns the whole model. Output format refers to the format selected to communicate the explanation (e.g. numerical, textual, visual, mixed...). Functioning refers to the process applied to extract information from an AI model to create an explanation (e.g. the perturbations method consists of providing to the AI model slight variations of a relevant input value to identify the importance of certain features on the model's prediction). Result refers to approaches that use the results of the AI model's prediction as the main input for the explanation. Other refers to other dimensions and particularly targets the approaches that aim to produce explanations according to a type of problem or a type of input data.

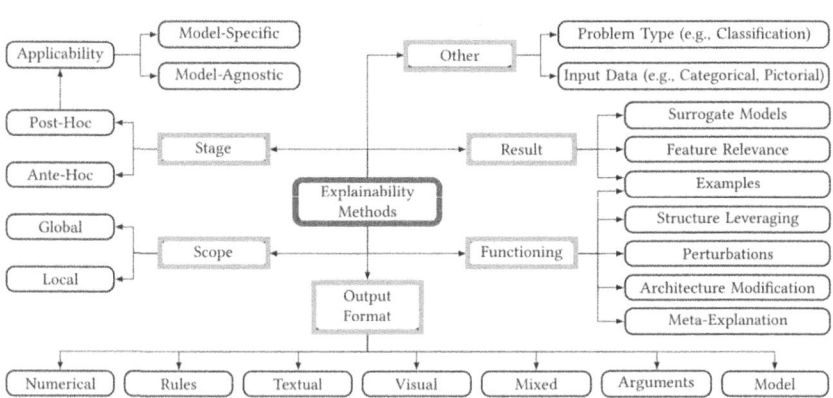

Fig. 2. Taxonomy of explainability methods (adapted from [14])

The explainability methods have several limitations. Even directly explainable models (ante-hoc) may need post-hoc explanations. They are so complex that they are black boxes for end-users and not fully understood by experts [15]. Post-hoc (or black box) explanations cannot be completely accurate as they are based on an AI model that is itself an interpretation of the original AI model (Rudin) [16]. Rudin even argues that we should not use the word "explanation" and use the term "summary statistics of predictions made by the original model" instead. Furthermore, explanations do not provide enough relevant information to be understandable by users [16], and they miss contextual information that may have an impact on the decision taken from the prediction [16].

Beyond not providing enough relevant information and missing contextual information, none of the explainability methods explicitly take into account user tasks with the AI model's prediction. Whereas, explanations have to be consistent with user tasks and have to support reaching the main user goals. They have to suit and integrate with the user tasks. In addition to consistency between explanation and user tasks, given the different types of explainability methods, the design of XAI in interactive systems requires ensuring compatibility between the type(s) of explanations provided and user tasks for each user goal. This means that the types of explanations have to be chosen to be the most relevant to user tasks, as well as to the possible contexts in which they could be provided. Another important point is that designing XAI for interactive systems requires assessing the risk that an explanation may trigger confusion and error for the next tasks. In that case, designers also have to analyze the impact of a possible misunderstanding of the explanation on the success of the user goals.

We have used this critical issue of explainability with AI technologies in this section to highlight the inherent complexity of the issues related to the deployment of AI technologies. Next section goes beyond this explainability issue and present the more global issue of certification of AI technologies as currently put forward by the European Union Aviation Safety Agency).

3 Certification of AI Technologies in Civil Aviation Domain

In 2020, the EASA (European Union Aviation Safety Agency) started the study of AI and its possible applications as a human centric approach to AI in aviation. This results in a roadmap [20] (updated in 2023 [21]) and the two first versions of guidelines for machine learning [19], pointing out that some issues remain to be addressed (cited from the foreword section of [20]): public confidence in AI-enabled aviation products, certification and approval of advanced automation, ethical dimension of AI, and processes, methods and standards to further improve the current level of air transport safety.

EASA proposes three levels of AI applications from assistance to human to AI-based decision making (see Fig. 3).

The following sections present how EASA planned to study AI technologies and present a focus on the guidance for the two first levels (the third is not studied yet).

3.1 Roadmap 2.0 to AI

In the second version of its roadmap [21], the EASA proposed the following timeline (see Fig. 4) as the result of both its planned deliverables and the forecasts of the main stakeholders in the large commercial air transport (CAT).

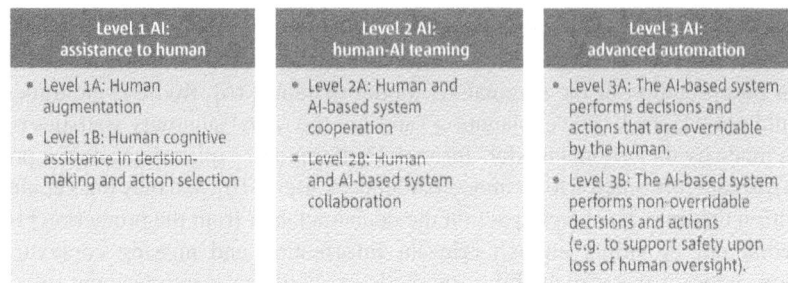

Fig. 3. EASA classification of AI applications from [21]

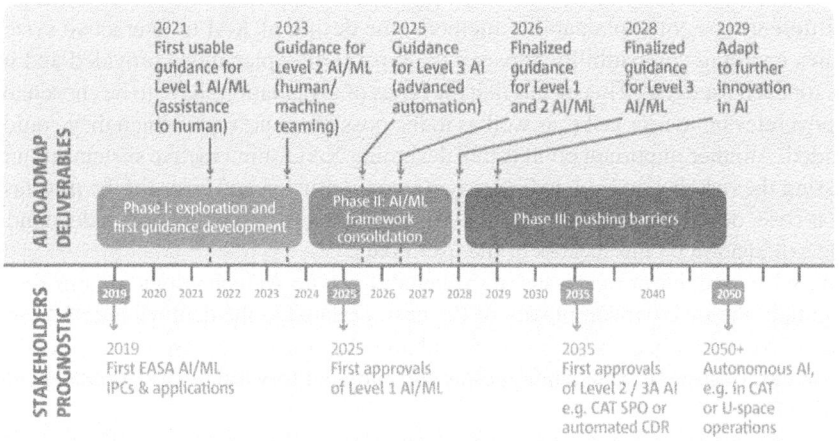

Fig. 4. EASA Roadmap 2.0 timeline from [21]

In this timeline the level 1 AI is foreseen for 2025, while level 2 AI (including more automated solutions to assist the pilot flying in extended minimal crew operations and single-pilot operations (SPO) is foreseen for around 2035 (it is not a formal target but a prognostic).

As stated in the roadmap document, the level 3 AI (advanced automation with ultimately no human supervision) is "most probably to be shifted even after 2050, considering the current state of the art in AI technologies".

Even for the main stakeholders of the domain, reaching levels of AI above the first one remains a bet because of the societal acceptability of AI. To face this difficulty, EASA planned three main phases (as shown on Fig. 4) which should output guidance (the first two are already available).

3.2 Guidance Level 1&2 to ML Applications

In 2023 (one year later than initially expected on the roadmap 1.0), EASA produced its first guidance for level 1&2 machine learning applications [19] as the basis for phase 2 of the roadmap. This document is a first set of objectives as an anticipation of the future guidance and requirements for safety-related machine learning applications.

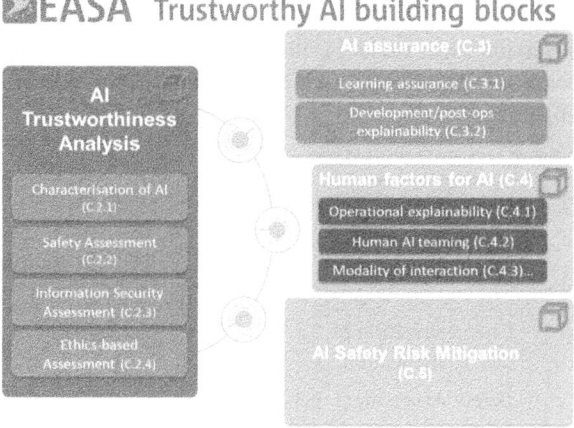

Fig. 5. EASA AI trustworthiness building blocks from [19]

As stated by EASA, the introduction of AI/ML technologies will have possible impacts on implementing rules (IRs), certification specifications (CSs), acceptable means of compliance (AMC) and guidance material (GM) and presents challenges (quoted from [19]):

- Adapting assurance frameworks to cover the specificities of identified AI techniques and address development errors in AI-based systems and their constituents.
- Dealing with the particular sources of uncertainties associated with the use of AI/ML technology.
- Creating a framework for data management, to address the correctness (bias mitigation) and completeness/representativeness of data sets used for the ML items training and their verification.
- Addressing model bias and variance trade-off in the various steps of ML processes.
- Elaborating pertinent guarantees on robustness and on absence of 'unintended function' in ML/DL applications.
- Coping with limits to human comprehension of the ML application behavior, considering their stochastic origin and ML model complexity.
- Managing shared operational authority in novel types of human-AI teaming.
- Managing the mitigation of residual risk in 'AI black box'. The expression 'black box' is a typical criticism oriented at AI/ML techniques, as the complexity and nature of ML models bring a level of opaqueness that make them look like unverifiable black boxes (unlike rule-based software).
- Enabling trust by end users.

The proposed set of objectives is divided into 4 parts (Trustworthiness analysis, AI assurance, Human-factors for AI and Safety risk mitigation), where the notion of proportionality allows objectives customization.

Figure 5 shows the framework proposed by EASA that is mainly based on the AI trustworthiness Analysis, which is mandatory for any AI application (left part of the figure). The depth of the guidance of the 3 other blocks may be adapted depending on the application (concept of proportionality represented by the potentiometers in the middle of the figure).

In its current version (see Table 1), this document proposes 118 objectives and 69 anticipated means of compliances (which are some of the potential strategies for achieving these objectives).

Table 1. Distribution of the objectives amongst the blocks of the EASA framework.

Framework block	Sub block	Numbers of Objectives	Anticipated Means of Compliance
AI trustworthiness analysis	Characterization of the AI application	7	6
	Safety Assessment	3	2
	Information Security Assessment	3	2
	Ethics based assessment	9	6
AI assurance	Learning Assurance	46	33
	Development & post-ops AI explainability	4	2
Human Factors for AI	AI operational explainability	13	7
	Human-AI teaming	13	3
	Modality of interaction	14	4
	Error management	4	2
	Failure Management	EASA proposes a matrix of responsibilities	
	Workload management	Topic under development	
	customization of human-AI interface	Topic under development	
AI safety risk mitigation	AI safety risk mitigation	2	2

Additionally, EASA proposes objectives modulations depending on the criticality of the targeted applications (the principle of proportionality).

3.3 Lessons Learned

When dealing with the criticality of AI applications, the EASA states that: "With the current state of knowledge of AI and ML technology, EASA anticipates a limitation on the validity of applications when AI/ML constituents include IDAL A or B / SWAL 1 or 2 / AL 1, 2 or 3 items (the highest level of criticality)" (in [19]).

This statement is in line with common practices in the field of safety critical systems engineering. For instance, if the domain of civil aviation, EASA defines five Development Assurance Levels (EASA CS-25 standard [22]) associated with their failure condition category and its description (summarized by Table 2).

Table 2. System Development Assurance Level for civil aircrafts

Development Assurance Level	Failure condition categories	Description of the failure conditions	Failure rate (failures/hour)
A	Catastrophic	Failure conditions that may cause a crash	Extremely improbable 10^{-9} + fail safe
B	Hazardous	Failure has a large negative impact or performance, or reduces the ability of crew to operate the plane	Extremely remote 10^{-7}
C	Major	Failure is significant, but has lesser impact than hazardous	Remote 10^{-5}
D	Minor	Failure is noticeable, but has lesser impact than Major	Probable 10^{-3}
E	No safety effect	No impact on dependability	Any range

According to this standard the first two rows of Table 2 (colored in grey) correspond to so-called "critical systems" while systems in the lower rows are called "non-critical". The standard DO-178C [17] states that implementing such critical systems requires achieving several objectives whose implementation is defined in the annexes of this standard. For instance, DO-333 annex [18] states that high level requirements should be expressed in temporal logic while low-level requirements should be expressed by state-based description techniques.

[23] recalls the classic means for the implementation of critical systems (see Fig. 6).

Accuracy reached with the classical approaches in the implementation of critical systems using formal methods and fault-tolerance mechanisms is still expected when AI technologies are introduced. For instance (quoted from [19]):

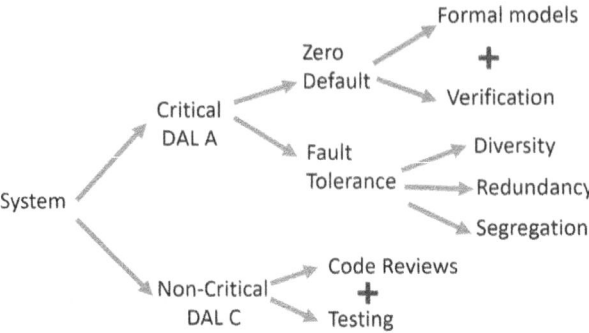

Fig. 6. Classical means for the implementation of critical systems in civil aviation.

- **Objective EXP-06**: "The applicant should present explanations to the end user in a clear and unambiguous form". **Anticipated means of compliance EXP-06**: The explanation provided should be presented in a way that is perceived correctly, can be comprehended in the context of the end user's task and supports the end user's ability to carry out the action intended to perform the tasks.
- **Objective EXP-13:** "The AI-based system should be able to deliver an indication of the degree of reliability of its output as part of the explanation based on actual measurements (e.g. monitoring) or on a quantification of the level of uncertainty". **Anticipated means of compliance EXP-13**: Assuming that the decisions, actions, or diagnoses provided by an AI-based system may not always be fully reliable, the AI-based system should compute a degree of reliability of its outputs. Such an indication should be part of the elements provided within the explanations when needed.

These two examples show the EASA expectation in terms of accuracy and highlight the quote from [19]: "Moreover, no assurance level reduction should be performed for items within AI/ML constituents". Next section presents two illustrative examples of AI-similar designs engineered with formal methods dedicated to interactive systems as presented in [27].

4 Two Illustrative Examples

In this section we propose two examples where developers could be led to AI technologies. The first example is presented with details: we firstly present its basic concepts and then discuss how AI technologies would "naturally" fit its development. We lastly present how it was actually designed. For lack of space, the second example is only introduced and could be fully presented in a longer version of this paper.

4.1 Fortunettes Interaction Technique

The origin of Fortunettes [1] is the need of providing feedforward about the future state of an application. This specific feedforward information display is triggered by the user on demand (when needed). In this approach, exploring the future may be seen as a four steps process:

- **Look at the present**, when the user explores visually the user interface elements;
- **Peek into the future**, when the user is considering performing an action;
- **Go to the future**, when the user confirms and actually executes the considered action;
- **Return to the present**, when the user is no longer considering the execution of that action.

The choice has been made of providing such feedforward at widget-level as it makes it easier to reuse for any widget-based application. Figure 7 shows an example of this kind of widget-level feedforward: when the user hovers over the Login button (that is currently enabled), the button Logout and the text box (that are currently disabled), show their future state in terms of availability (the button Logout and the textbox will become enabled if the user clicks the button Login, while the button Login will become disabled). With this information, the user knows that to enable the Logout button, the Login button must be pressed first.

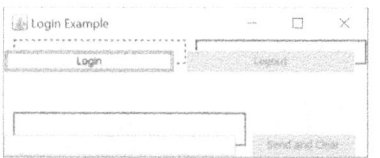

Fig. 7. Illustration of the Fortunettes concept using the case study.

The main idea of Fortunettes is to provide the user with an answer to "What will happen if I do that?", by presenting what the result of the user action will be, before the action is actually performed. It thus requires the widgets to be able to present their future state in addition to their current state. As presented in [1], the user interface of the application presented in Fig. 7 is the following:

- The application is composed of four widgets (the three buttons and a text-box),
- The current state of the widgets is displayed on the forefront, the Login button is enabled, while Logout and Send and Clear ones and the textbox are disabled.
- In order to present the feedforward information, users have to hover over the widget of interest. In Fig. 7, the Login button is hovered and the background display of each widget presents the feedforward information showing the state of the application if the user clicks on the login button. Current feed-forward display tells the user that Login button will be disabled, the textbox will become enabled, Logout will become enabled and Send and Clear will remain disabled. Indeed, as the status of Send and Clear will remain the same, no additional feedforward display is presented.

Why AI Technologies Would be Perceived as a Good Option?
The Fortunettes step "peek into the future" is a good candidate for AI technologies. For instance, A* algorithms are a way to explore all possible futures without reaching these futures using a heuristic function to estimate the distance to the objective. As most software developers are trained in AI teaching units to produce and use such algorithms, they will naturally think at using them for implementing fortunettes widgets.

Formal Based Engineering of Fortunettes without AI Technologies

As stated in the introduction of this section, engineering an application with feedforward capabilities requires to handle extra interaction events (at least three, depending on the widget type). These events allow the user to **peek into the future**, to **go to the future** or to **return to the present**, without affecting the standard behavior of the application, as the objective is to enhance the application (with feedforward) and not to change it. This design choice impacts the modelling of feedforward behavior:

- The feedforward behavior of any application is modelled as an independent object that embeds the standard behavior (as a copy), making it fully compatible with the original application behavior. This Petri net model is called the **Fortune Net** as it allows users to look into the future of the application.
- For any event handling within the standard behavior, the feedforward behavior embeds a pattern described in Petri nets (a set of places and connected transitions) that models the exploration of the future states. The important aspect in this modelling principle is that we exploit the behavior of the application to forecast the future states of the application if the user decides to use feedforward function.

To illustrate these two points, we use an excerpt of the complete behavior that only concerns the Login action on the user interface (as shown by Fig. 8).

Fig. 8. Excerpt from the Petri net model of the standard behavior of the application: event handling of the login action. In the transition, the text on the left describes the name of the transition while the text on the right describes the name of the event associated the transition.

In Fig. 8, the login transition is the event handler for an event called loginPerformed that represents the use of the button Login. When fired, this transition moves the token from place LoggedOut (1.) to place LoggedIn, setting the state of the application to the new state following the execution of the login (code not represented here).

The three base actions (peek into the future, go to the future and return to the present) are represented as three extra event handlers, as shown on Fig. 9, where event handlers {FloginPerformed, UFloginPerformed, InFloginPerformed} are generated from the event handler loginPerformed.

On Fig. 9, transition f1login (event handler for FloginPerformed) represents the action of peeking into the future of the action login. Basically, it behaves in the same way as the original action (put a token in place LoggedIn) while the standard behavior is still in state LoggedOut. It additionally puts a token in place flogin that represents the entering in feedforward mode (a dedicated rendering may occur).

There are then two possibilities:

- The user decides to really perform the login action (using the login button), producing two events: loginPerformed handled by the standard behavior (making it going to the state LoggedIn) and InFloginPerformed handled by the

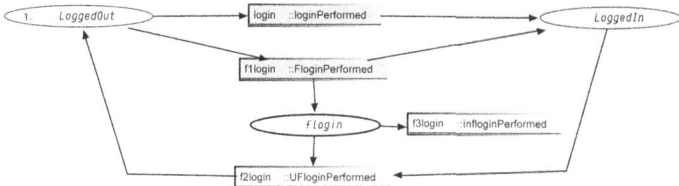

Fig. 9. Extracted from the feedforward behavior of the application: event handling of the login action and peek into its future.

feedforward behavior (discarding the token in place flogin, while the token in place LoggedIn does not move, placing it in the same state as the standard behavior).
- The user decides to not perform the login action producing an event UFloginPerformed. The standard behavior remains in the same state while in the feedforward behavior, the tokens from places LoggedIn and flogin are removed and a token goes back to the place LoggedOut, making it return to its previous state (leaving the feedforward mode).

This pattern is particularly efficient when describing a feedforward behavior for events that do not handle values or when the widgets are simple such as button. For more complex events, or when the underlying widgets are more complicated, this pattern has to be modified/extended:

- When values are handled by the action of the widget, it is not always possible to peek into the future of these values. One possible improvement is to proceed in two steps. When entering the feedforward mode, an envisioned value must be produced (decided at design time for instance) and when the user really performs the action, a substitution must be done between the envisioned value and the real value. In the feedforward behavior, this can be done by moving tokens (if it was the case in the login example, the first token put in place LoggedIn by transition f1login would have a design time envisioned value, and when f3login would be fired, this token would have been removed and replaced by one holding the correct value).
- When the widget is more complex (in our case, the complexity is related to the event production), extra event handlers may be introduced. For instance, when using a classical textbox, one may be interested by the end of the text edition (validation) and not by the whole process of typing in the text. In this case, in the standard behavior of the application, the only handled event would be the last one (for instance, the event actionPerformed of the JTextField in Java Swing). On the feedforward behavior side, any text change may be relevant to allow the rendering of text filtering.

Fortunettes requires enhancing widgets with extra means to allow rendering feedforward states and to trigger dedicated events. In our implementation using Java Swing widgets, we embed them within a specialized decorator, but there are many other implementation options at widget level or at application level.

When the design and development of the application reaches a state in which it can be executed, we can test and validate it using the ICO runtime environment. Figure 10. Depicts how the runtime environment comes into play.

User events are transmitted to both the standard ICO model as well as the ICO feedforward model. In this case a click-event triggers an action and progresses both the standard ICO model as well as the ICO feedforward model. A hover, on the other hand, triggers the ICO feedforward net thus computes and presents what the potential outcome is if one would confirm the action with a click. Notice the trigger to show feedforward, i.e. a hover action, can be replaced by any other actions supported by ICO.

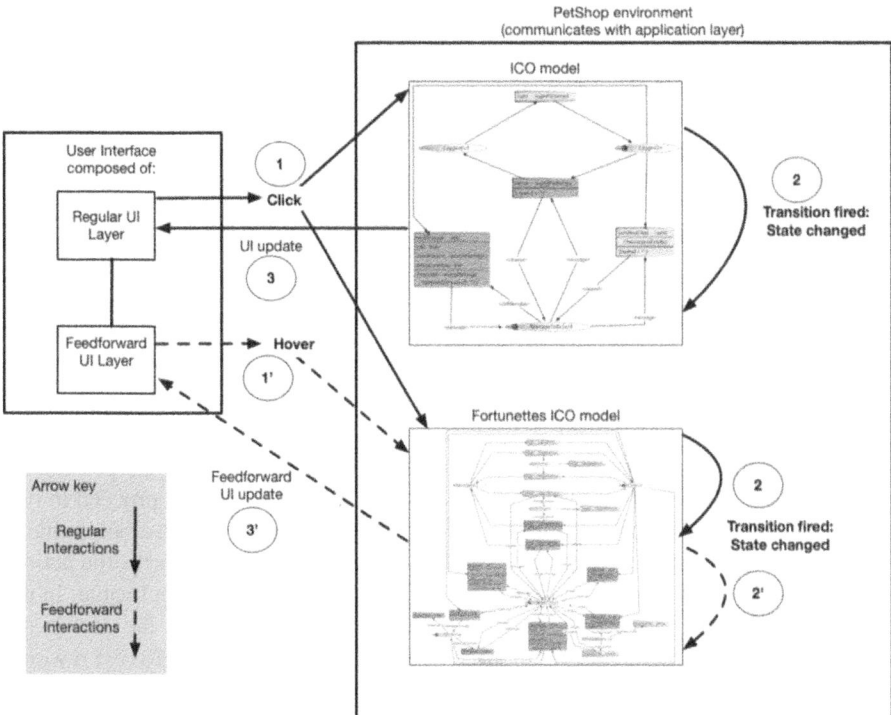

Fig. 10. Runtime architecture of ICOs and ICOs fortunettes models in PetShop environment, where a hover action is used to trigger feedforward. Sequence 1-2-3 presents what happens when an actual interaction is triggered. Sequence 1'-2'-3' presents what happens when feedforward is triggered.

4.2 A Recommender System for an Aircraft Flight Warning System

This section presents a system from large commercial aircraft that could be a good candidate to become a recommender system in future programs. The left part of Fig. 11, presents the current Flight Warning System (FWS) of the aircraft A350. This FWS is part of the ECAM (Electronic Centralized Aircraft Monitor) and system provides support to the pilot to manage abnormal situation, by displaying alerts about system failures and procedures that have to be completed by the pilot to manage the detected warning. The current FWS displays only the procedure associated to the top ranked alarm (predefined absolute ranking) and a procedure describes a list of recovery actions to resolve the associated alarm. The pilot has to perform sequentially the recovery actions.

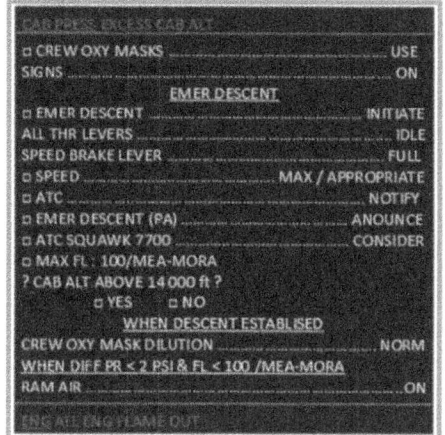 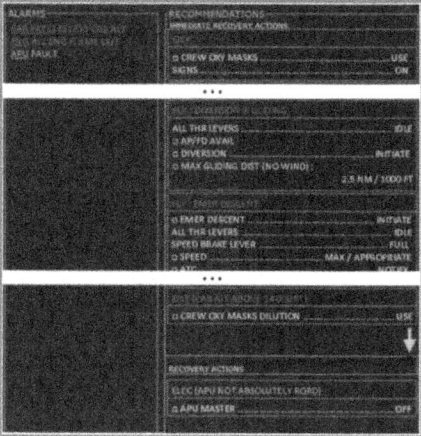

Fig. 11. On the left) the A350 Flight Warning System (FWS). On the right) the Recommender Flight Warning System (REC-FWS)

The right part of Fig. 11 presents an FWS based on recommendations (REC-FWS). The REC-FWS presents sets of recommended actions from which the pilot will choose and select according to the pilots knowledge and the context only known by the pilot (such as a loss of consciousness of a passenger). The REC-FWS does not present procedures but sets of recovery actions to perform a higher-level user task such as performing an emergency descent.

Due to space constraints, we do not present here how such recommender is designed and implemented but as explained in [5] machine learning technologies cannot be deployed in civil aviation.

Why AI Technologies Would be Perceived as a Good Option?

As the REC-FWS is a recommender system, it seems obvious that AI technologies can meet its requirements. We all know existing recommender systems (Amazon, Netflix…) which claim machine learning [12] as underlying technologies despite the fact that HCI researchers also know that it is not beneficial for the end user [12].

A Software Architecture for Recommender Systems Without AI Technologies

Figure 12 presents a generic architecture for recommender systems. This architecture may be divided into two distinct data flows.

The first one starts at the bottom of Fig. 12 where sources are used to feed the database on which recommendations are built. These data are pre-processed to produce items that are analyzed to build recommended items. These recommended items lead to both recommendations and related explanations that can be presented to a user through a dedicated user interface.

The second data flow comes with the user interaction with the recommendations (top right part of Fig. 12) that is reintroduced in the first data flow in two ways: update of the user profile and production of new recommendations that are then possibly presented back to the user. The basic bricks of this architecture do not necessarily require AI technologies but employing them would speed up the development process at the expense of dependability which is not acceptable in the application domain considered.

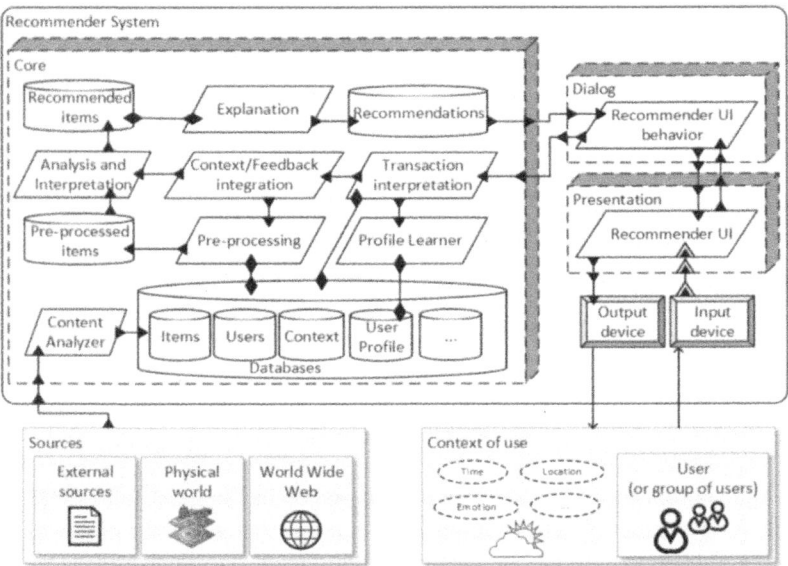

Fig. 12. A Generic software architecture for recommender systems not requiring AI technologies

5 Conclusion

This paper highlighted the fact that complex interactive computing systems can be engineered without integrating AI technologies, even though, at first sight, users and developers might have believed differently. Indeed, designs embedding automation of operators' tasks (such as aircraft autopilots or Flight Management Systems) are nowadays perceived as AI-similar designs and thus believed to embed AI technologies.

We have demonstrated, using an interactive technique example and a recommender system in the context of aviation, that such AI-similar designs can be implemented using formal engineering techniques for interactive computing systems as [28] (or [31] for multimodal interactions) which uses Petri nets or [29] which uses Event-B. Using such technologies implement designs that are AI-similar but provide dependability guarantees not available with AI technologies.

To go beyond the work presented here we have identified the following three challenges:

- The need to identify criteria to assess where and when AI technologies are needed and thus must be integrated into software engineering approaches;
- The need to provide systematic means to integrate such technologies into complex software systems ensuring reliability and dependability to the required level;
- The need to define systematic means to assess benefits and drawbacks on the integration of these technologies in terms of usability, reliability, dependability and eventually safety.

Addressing those challenges would support software engineers in understanding, in the early phases of the development process, which of the candidate technologies have to be considered and to what extend they meet the needs. This key issue of feasibility is carefully addressed in tasks management on computing resources (as in [30]) and similar paths could be followed.

References

1. Solom, O.: The rise of 'pseudo-AI': how tech firms quietly use humans to do bots' work. The guardian, Fri 6 Jul 2018 08.01 BST. https://www.theguardian.com/technology/2018/jul/06/artificial-intelligence-ai-humans-bots-tech-companies
2. Lubars, B., Tan, C.: Ask not what AI can do, but what AI should do: towards a framework of task delegability. In: 33rd Conference on Neural Information Processing Systems (NeurIPS 2019), Vancouver, Canada (2019)
3. Coppers, S., Luyten, K., Vanacken, D., Navarre, D., Palanque, P., Gris, C.: Fortunettes: feedforward about the future state of GUI widgets. In: Proceedings of the ACM on Human-Computer Interaction, vol. 3. ACM SIGCHI (2019)
4. Navarre, D., Palanque, P., Coppers, S., Luyten, K., Vanacken, D.: Fortune nets for fortunettes: formal, petri nets-based, engineering of feedforward for GUI widgets. In: Sekerinski, E. (ed.) Formal Methods. Lecture Notes in Computer Science, vol. 12232, pp. 503–519. Springer, Cham (2020). https://doi.org/10.1007/978-3-030-54994-7_36
5. EASA Artificial Intelligence Roadmap 1.0: A human-centric approach to AI in aviation. https://www.easa.europa.eu/en/document-library/general-publications/easa-artificial-intelligence-roadmap-10
6. Navarre, D., Palanque, P., Coppers, S., Luyten, K., Vanacken, D.: Model-based engineering of feedforward usability function for GUI widgets. Interact. Comput. **33**(1), 73–91 (2021). https://doi.org/10.1093/iwcomp/iwab014
7. Bouzekri, E., Martinie, C., Palanque, P., Atwood, K., Gris, C.: Should I add recommendations to my warning system? The RCRAFT framework can answer this and other questions about supporting the assessment of automation designs. In: Ardito, C., et al. (eds.) Human-Computer Interaction -INTERACT 2021. Lecture Notes in Computer Science, vol. 12395, pp. 405–429. Springer, Heidelberg (2021). https://doi.org/10.1007/978-3-030-85610-6_24
8. Bouzekri, E., et al.: Engineering issues related to the development of a recommender system in a critical context: application to interactive cockpits. Int. J. Hum. Comput. Stud. **121**, 122–141 (2019)
9. Bouzekri E., et al.: A list of pre-requisites to make recommender systems deployable in critical context. In: EnCHIReS@EICS. CEUR Proceedings (2017)
10. Fayollas, C., Martinie, C., Palanque, P.A., Barboni, E., Deleris, Y.: What can be learnt from engineering safety critical partly-autonomous systems when engineering recommender systems. In: EnCHIReS@EICS, pp. 14–25 (2016)

11. Gomez-Uribe, C., Hunt, N.: The netflix recommender system: algorithms, business value, and innovation. ACM Trans. Manage. Inf. Syst. **6**(4), 19 p. Article 13 (2016). https://doi.org/10.1145/2843948
12. Schaffner, B., Stefanescu, A., Campili, O., Chetty, M.: Don't let Netflix drive the bus: user's sense of agency over time and content choice on Netflix. In: Proceedings of ACM Human-Computer Interaction. CSCW1, vol. 7, 32 p. Article 128 (2023). https://doi.org/10.1145/3579604
13. Gilpin, L.H., Bau, D., Yuan, B.Z., Bajwa, A., Specter, M., Kagal, L.: Explaining explanations: an overview of interpretability of machine learning. In: 2018 IEEE 5th International Conference on Data Science and Advanced Analytics (DSAA), Turin, Italy, pp. 80–89 (2018). https://doi.org/10.1109/DSAA.2018.00018
14. Speith, T.: A review of taxonomies of explainable Artificial Intelligence (XAI) methods. In: Proceedings of the 2022 ACM Conference on Fairness, Accountability, and Transparency (FAccT '22), pp. 2239–2250. Association for Computing Machinery, New York (2022). https://doi.org/10.1145/3531146.3534639
15. Burrell, J.: How the machine 'thinks': understanding opacity in machine learning algorithms. Big Data Soc. **3**(1), 1–12 (2016). https://doi.org/10.1177/2053951715622512
16. Rudin, C.: Stop explaining black box machine learning models for high stakes decisions and use interpretable models instead. Nat. Mach. Intell. **1**(5), 206–215 (2019). https://doi.org/10.1038/s42256-019-0048-x
17. DO-178C / ED-12C, Software considerations in airborne systems and equipment certification, published by RTCA and EUROCAE (2012)
18. DO-333 Formal Methods Supplement to DO-178C and DO-278A, published by RTCA and EUROCAE, 13 December 2011
19. EASA, 'Concept Paper: Guidance for Level 1 & 2 machine learning applications—Proposed Issue 02', European Union Aviation Safety Agency (EASA), Cologne (2023)
20. EASA, 'EASA Artificial Intelligence Roadmap 1.0', European Union Aviation Safety Agency (EASA), Cologne (2020)
21. EASA, 'EASA Artificial Intelligence Roadmap 2.0', European Union Aviation Safety Agency (EASA), Cologne (2023)
22. CS-25 – Amendment 17 - Certification Specifications and Acceptable Means of Compliance for Large Aeroplanes. EASA (2015)
23. Fayollas, C., Martinie, C., Palanque, P., Deleris, Y., Fabre, J.-C., Navarre, D.: An approach for assessing the impact of dependability on usability: application to interactive cockpits. In: European Dependable Computing Conference, pp. 198–209 (2014)
24. Welsh, M.: The end of programming. Commun. ACM **66**(1), 34–35 (2023). https://doi.org/10.1145/3570220
25. Greengard, S.: AI rewrites coding. Commun. ACM **66**(4), 12–14 (2023). https://doi.org/10.1145/3583083
26. Meyer, B.: AI does not help programmers. In: Blog@acm (2023). https://cacm.acm.org/blogs/blog-cacm/273577-ai-does-not-help-programmers/fulltext. Accessed 3 June 2023
27. Campos, J.C., Fayollas, C., Harrison, M.D., Martinie, C., Masci, P., Palanque, P.: supporting the analysis of safety critical user interfaces: an exploration of three formal tools. ACM Trans. Comput.-Hum. Interact. **27**(5), 48 p. Article 35 (2020). https://doi.org/10.1145/3404199
28. Hamon, A., Palanque, P., Silva, J.L., Deleris, Y., Barboni, E.: Formal description of multi-touch interactions. In: Proceedings of the 5th ACM SIGCHI Symposium on Engineering Interactive Computing Systems (EICS '13), pp. 207–216. ACM (2013). https://doi.org/10.1145/2494603.2480311
29. Mendil, I., Aït-Ameur, Y., Singh, N.K., Dupont, G., Méry, D., Palanque, P.: Formal domain-driven system development in Event-B: application to interactive critical systems. J. Syst. Archit. **135**, 102798. ISSN 1383-7621 (2023). https://doi.org/10.1016/j.sysarc.2022.102798

30. Sung Chwa, H., Lee, J.: Tight necessary feasibility analysis for recurring real-time tasks on a multiprocessor. J. Syst. Archit. **135**, 102808 (2023). ISSN 1383-7621. https://doi.org/10.1016/j.sysarc.2022.102808
31. Palanque, P., Schyn, A.: A model-based approach for engineering multimodal interactive systems. In: IFIP TC 13 conference INTERACT 2003 (2003). https://hal.science/hal-03664744v1

Explaining Through the Right Reasoning Style: Lessons Learnt

Lucio Davide Spano[1](✉) and Federico Maria Cau[2]

[1] University of Cagliari, Cagliari, Italy
davide.spano@unica.it
[2] Maastricht University, Maastricht, The Netherlands
federico.cau@maastrichtuniversity.nl

Abstract. Current eXplainable Artificial Intelligence (XAI) techniques assist individuals in interpreting AI recommendations. However, research primarily focuses on assessing users' comprehension of explanations, neglecting important factors influencing decision support, such as whether the explanation uses the correct reasoning style to help the user understand the AI's advice. In the last two years, our research aimed to fill this gap by examining the effects of factors such as user uncertainty, AI correctness, and the interplay between AI confidence and explanation logic styles in classification tasks. In this paper, we summarise the lesson learnt from this research and discuss its impact on the engineering of AI-based decision support systems.

Keywords: Explainable AI · User uncertainty · AI uncertainty · AI correctness · Explanations · Logical Reasoning · Inductive · Deductive · Abductive

1 Introduction

Over the past decade, significant progress has been made in eXplainable Artificial Intelligence (XAI) research, introducing various explanation techniques [7,10–12]. Research studies on these techniques usually focus on their level of interpretability, i.e., the extent to which a human can comprehend the reasons behind decisions made by AI methods [9]. However, the final goal of a decision support system is fostering correct decisions (*task performance*). In this regard, it is also interesting to examine whether the user follows AI advice in decision-making (*agreement*) and which examines pieces of information in the XAI interface that influenced the final decision (*reliance*).

We assessed our theory on the factors influencing this process in two different research studies [3,4]. To describe how users make the final decision, we should consider their confidence in their judgement after analysing the decision instance. When the users have domain knowledge and their confidence is high, they may rely only on this for making the final decision. However, it happens that users are confident about a wrong decision. In general, such a situation occurs when

i) there is a lack of information (e.g., the instance does not sufficiently describes the situation) and ii) when the observation of the instance can have diverse interpretations among different user groups and at least one of them is wrong. The second cause is the most relevant and appropriate to cover in a decision support system, which should guide users in making the wrong interpretation to revise their judgement.

Considering this, we expected that user confidence would be a poor predictor for the final decision correctness. We suppose more appropriate the *user's uncertainty* as the distribution of users' accuracy per instance within a user population, which aims to quantify the difficulty level in a subset of observations based on the users' performance.

The role of AI in the process requires more modelling variables. If users consider its suggestion and have some domain knowledge, the *AI correctness* should influence the final outcome. However, suppose the user is highly uncertain about which decision to make, for instance, when their domain knowledge is low. In that case, it is also likely that users will rely on AI, no matter whether the outcome is correct or not.

But how can users assess the AI's prediction? Explanations have this role in XAI interface: they should convey why the AI is suggesting a certain decision. The strength of this "arguments" is related to the *AI confidence* (or uncertainty): the more confident the AI, the more the explanations should be able to let the users interpret the rationale behind the AI's decision. This does not mean that the explanation should persuade the user to follow the AI: it is possible that if the AI is confident about a wrong decision, the explanation persuades the user to reject the suggestion. If so, the task performance should increase, even if there is no agreement between users and AI.

However, we expect that some explanations have different effectiveness in communicating the AI's confidence to the users. Among all the characteristics related to explanations, we focused on the *reasoning style* they use. Inductive reasoning involves deriving a general conclusion from a specific set of observations. Abductive reasoning, on the other hand, starts with incomplete observations and seeks to find the most plausible explanation. Deductive reasoning begins with general rules and uses them to reach a specific and logical conclusion. We expect that such reasoning styles interact with AI confidence, and their effect depends on the considered domain. Before our studies, only a few studies examined the effects of presenting explanations using different logical reasoning styles on users. Buçinca et al. [2] briefly discussed the impact of inductive and deductive reasoning explanations on users in a nutrition-related scenario. These explanations were designed through example-based explanations and general rules generated by a simulated AI. Another relevant article that explored different explanation styles, aligning with Pierce's theory, is the work by Van Der Waa et al. [14].

2 Result Summary

Until now, we assessed our hypothesis on three different domains: image recognition, text sentiment identification and stock market predictions (see Table 1).

In the first two domains, users have a good knowledge of the domain, and we focused on two different data types for representing the instance. In the third one, we selected users needing more domain knowledge.

Image Recognition. The test consisted of recognising digits from the MNIST dataset [6]. We selected two instances for all the combinations of the controlled variables: user uncertainty (high and low), AI confidence (high and low), AI correctness (right and wrong) and explanation style (no explanation, inductive, abductive, and deductive) A visual summary of the explanations styles is available in Fig. 2, top part. To generate inductive explanations, we utilise the weights of the AlexNet model and map them onto a k-NN (k-Nearest Neighbors) model to identify the three closest neighbours. We employ Grad-CAM [11], a technique that highlights important regions in an image for abductive explanations. Lastly, for deductive explanations, we leverage an encoder-decoder architecture based on the work of Vinyals et al. [15] to generate natural language sentences. In the study participated 659 users, each one recognising a single digit.

For the reliance, the findings indicate that when *user uncertainty* is low, users tend to rely more on the image itself (ranked first) rather than the AI prediction and explanation type, which aligns with our expectations. When the AI makes correct predictions (*AI correctness* is right), users tend to rely on the AI prediction as the second source of information and the explanation as the third source. When the AI makes incorrect predictions (*AI correctness* is wrong), the explanation becomes the second source of information, while the AI prediction falls to the third position.

For the task performance, a low *user uncertainty* and correct *AI correctness* have a positive impact on *Task Performance*, as expected. Additionally, it was found that inductive and deductive *explanation styles* have a negative impact on performance, particularly when *AI confidence* is high. This effect is more pronounced when the AI prediction is incorrect.

For the agreement, we observed that the *user uncertainty* had no significant effect. Instead, the *AI correctness* positively affected it. Finally, the interaction between *AI confidence* and *explanation style* was significant. In particular, inductive explanations increase the agreement for both a high or low-confident AI, while abductive explanations discourage users from accepting the AI suggestion when the confidence is high.

Text Sentiment. The evaluation setup was the same as the previous one. This time we considered instances of the Yelp Review dataset [1]. Figure 2, middle part, shows a visual summary of the different explanation styles for this task. As for the explanation styles, the inductive is generated using the three nearest neighbours obtained by mapping the prediction model to a k-NN. For the abductive and deductive explanations, we employ the information and visualisations provided by LIME [10], the former by highlighting the words that influenced the AI's decision indicating the importance of each word for the prediction through the opacity of its background colour, the latter visualising the top ten most significant words for each sentiment, and their weights in supporting or opposing the assignment to a specific class.

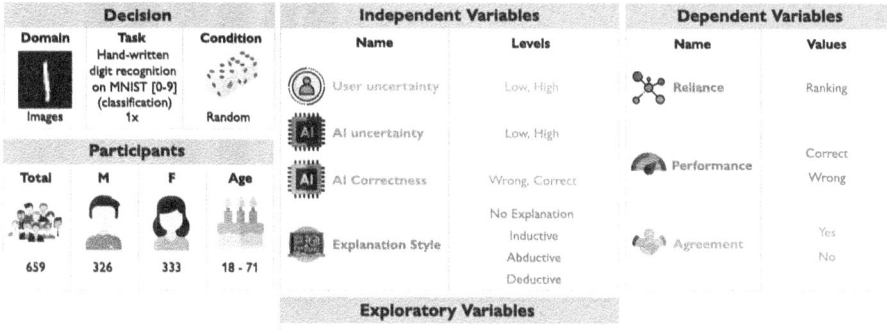
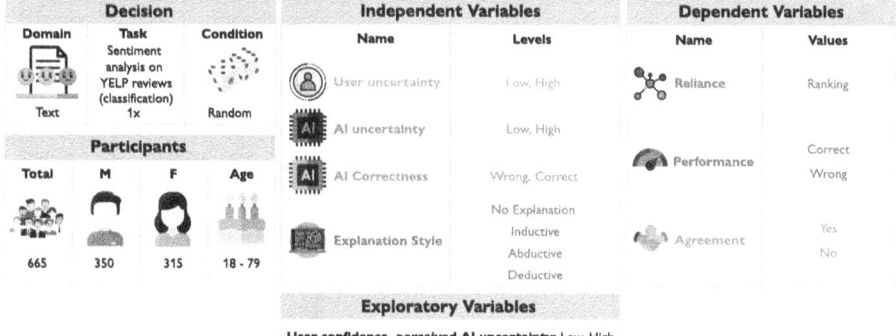
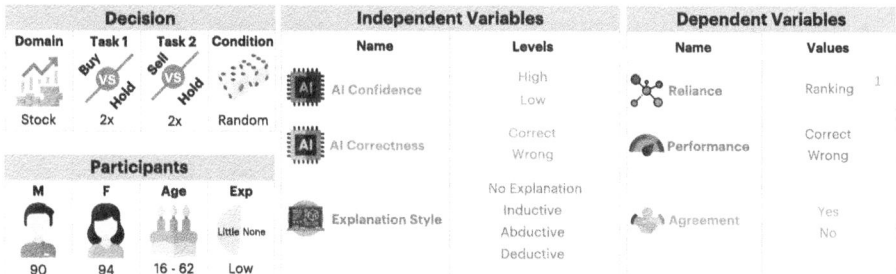

Fig. 1. Visual summary of the three studies discussed in this paper. The first one included a single image recognition task, the second was about text-sentiment recognition, and the third one included two buy-or-hold and two sell-or-hold tasks in the stock market domain.

Study 1: Image Data

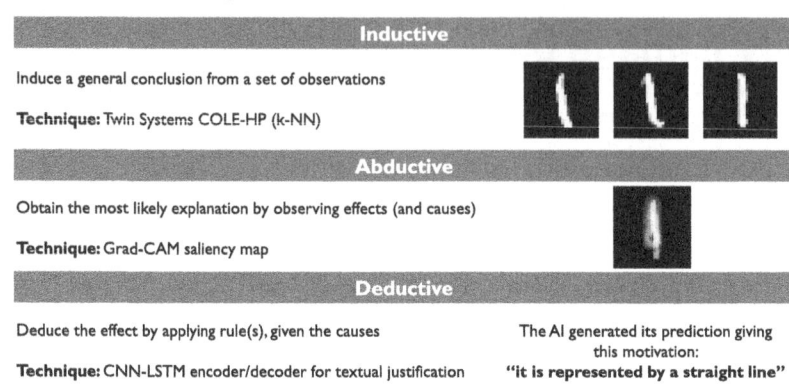

Study 2: Text Data

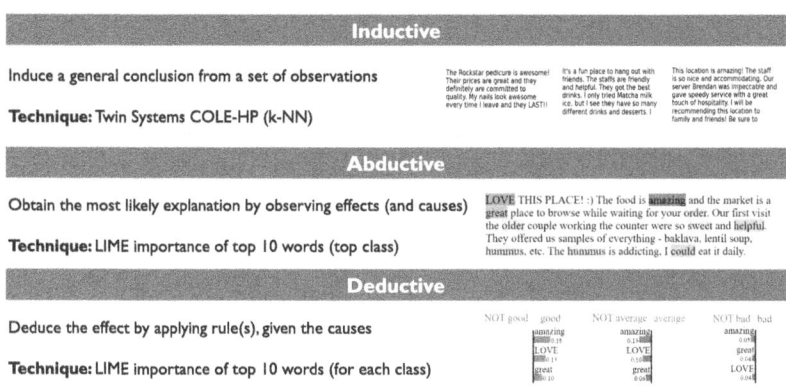

Study 3: Stock Market

Fig. 2. Visual summary of the three logical reasoning explanations (i.e., inductive, abductive, and deductive) for the three studies discussed in this paper: (1) image recognition task, (2) text-sentiment recognition, and (3) buy-or-hold and sell-or-hold tasks in the stock market trading task.

Table 1. Hypotheses results summary for image, text, and stock domains.

Hypotheses	Image	Text
H1: *Reliance*		
H1a: When the *User Uncertainty* is low, the user will rely more on the instance rather than on AI prediction and the explanation.	✓	✓
H1b: When the *User Uncertainty* is high, the user will rely more on the instance and the explanation rather than on the AI prediction.	✗	✓
H2: *Task Performance*		
H2a: Users will achieve a higher *Task Performance* with a low *User Uncertainty* than a high one.	✓	✓
H2b: Users will achieve a higher *Task Performance* with *correct AI* predictions than with wrong ones.	✓	✓
H2c: *AI uncertainty* moderates the effect of the *Explanation Styles* on the *Task Performance* (positively or negatively).	✓	✗
H3: *Agreement*		
H3a: Users will achieve a higher *Agreement* with a high *User Uncertainty* than a low one.	✗	✗
H3b: Users will achieve a higher *Agreement* with *correct AI* predictions than with wrong ones.	✓	✗
H3c: *AI uncertainty* moderates the effect of the *Explanation Styles* on the *Agreement* (positively or negatively).	✓	✓
Hypotheses	**Stock**	
H1: *Reliance*		
H1a: When the *AI Confidence* is high, the user will primarily rely on the charts with indicators or the AI prediction, then on the explanation.	✓	
H1b: When the *AI Confidence* is low, the user will primarily rely on the charts with indicators, then on the explanation or the AI prediction.	✗	
H2: *Task Performance*		
H2a: When the *AI Confidence* is high, abductive *Explanation Style* leads to a higher *Task Performance* if compared against the inductive.	✓	
H2b: When the *AI Confidence* is high, deductive *Explanation Styles* leads to a higher *Task Performance* if compared against the inductive.	✓	
H3: *Agreement*		
H3a: When then *AI Confidence* is high, abductive *Explanation Style* leads to a higher *Agreement* with *correct AI* predictions.	✓	
H3b: When then *AI Confidence* is high, deductive *Explanation Style* leads to a higher *Agreement* with *correct AI* predictions.	✓	

The findings suggest that when *user uncertainty* is low in the text domain, users tend to rely primarily on the text itself rather than the AI prediction and explanation. When it is high, users still prioritise the text but consider the explanation as the second source of information. This ranking pattern persists regardless of the *AI correctness*.

For the performance, both low *user uncertainty* and the correct level of *AI correctness* positively support our hypothesis. However, no significant interactions were observed between the factors of *AI confidence* and *Explanation Style*, indicating that our hypothesis was not supported in this context.

For the agreement, we have a positive impact when the *user uncertainty* is low, supporting our hypothesis. The *AI Correctness* factor does not significantly influence the agreement. Furthermore, a high *AI confidence* has a negative effect on the agreement with abductive explanations. When the *AI confidence* is low, all explanation styles increase the agreement compared to the no-explanation condition.

Stock Market. In this study, we focused on users having low domain expertise. We recruited 250 participants, each making four decisions of two types: i) buy or hold, supposing the user does not already own any stock, and ii) sell or hold, supposing they already own stocks. This time, the controlled variables are AI confidence, AI correctness and explanation style since the user uncertainty would be high for any decision, given the specific domain. We used a Random Forest classifier for the AI prediction. Inductive explanations were retrieved using the k-NN algorithm inside MACEst. Abductive explanations are based on the SHapley Additive exPlanations (SHAP) framework [7], highlighting the most relevant features in the stock descriptions for making the decision, while deductive used CHIRPS [5] for generating a set of rules applied for making the decision. The different explanations styles are depicted in Fig. 2, bottom part.

3 Lesson Learnt

When Users Have Domain Knowledge, They Are Overconfident about Their Decisions. Our findings in the image and text domain show a clear presence of user overconfidence in decision-making, i.e. the users rely too much on their decision without considering the AI's suggestions. This observation is supported by the results of the reliance analysis, which consistently indicate the instance as the most influential factor, regardless of the level of *user uncertainty*. This overconfidence in the instance choice helps explain the significance of *user uncertainty* in the performance, suggesting that users' judgement regarding the selected instance has a high impact on the final decision. These findings align with our initial expectations before conducting the studies.

Inductive Explanations. The structure of inductive explanations, which present similar instances of the same predicted class, creates a shared similarity concept between users and the AI, leading to an increased agreement even when the prediction is wrong in the image and text domains. They consistently persuade users to follow the AI suggestion. In addition, they have a negative effect on performance in image recognition when *AI confidence* is high. However, they do not significantly impact the performance in other conditions. Inductive explanations consistently persuade users to follow the AI suggestion in both image and text domains regardless of *AI correctness*. In contrast, stock prediction tasks involve cognitive effort in comparing indicators and identifying trends. In this context, employing an inductive inference process for explaining the AI prediction may not be optimal as it would increase the user's cognitive load, leading to the disregard of the explanation.

Abductive Explanations. Abductive explanations in the text and image domains do not significantly affect the performance but have a mixed effect on the agreement, depending on the *AI confidence*. When such confidence is high, abductive explanations lead to less agreement, while when the *AI confidence* is low, the agreement increased. This pattern is consistent in both the image and text domains. Abductive explanations seem to cause disagreement with a confident AI but foster agreement with an uncertain AI. This finding challenges the expectations of interface designers, and further research is needed to understand this phenomenon better. In contrast, in the stock prediction task, the abductive explanations provide a good support in identifying the elements of the instance description contributing to the AI prediction, facilitating the acceptance or the rejection of the AI's suggestion.

Deductive Explanations. Deductive explanations in the text and image domains have a negative effect when the *AI confidence* is high, while they do not affect the agreement. When *AI confidence* is low, they lead to less agreement in image data but they increase *agreement* in text data. While guiding how to use them in the general case is difficult, they do not seem the best option in these domains, considering their alternating effects and the effort required for producing them. In the stock market domains instead, the effort is worth it, considering that they improve the performance and the agreement when the AI is correct, similar to the abductive style.

The Role of AI Correctness. Providing the users with correct AI-based suggestions had a different impact. It was significant for image and text sentiment recognition, while stock market prediction did not change the users' performance. The high uncertainty associated with the stock market prediction task, coupled with the lack of domain expertise among study participants, made them equally likely to accept or reject correct and wrong AI predictions. Therefore, in this context of high uncertainty, the correctness of the AI predictions does not account for the observed over or under-reliance phenomenon reported in previous studies that motivated our work. Instead, we identified the explanation style and the AI confidence as relevant factors for users to appropriately consider the AI suggestion. Our findings suggest that for guiding non-experts through AI support, it may be more crucial to accurately estimate and communicate confidence in predictions using specific explanation styles such as abductive and deductive approaches.

Limitations. To carry out the three studies, we needed to make different decisions that, on the one hand, made the study feasible but, on the other hand, may limit the generalisability of the results. Examples of such choices are the decision domain, the selected instances, the decision task description and scenarios, the classification models, and other technical details such as the indicators used for the stock assessment in the last study. All choices were made thoughtfully, balancing the availability of state-of-the-art AI models and the techniques for providing explanations, together with representative tasks and the study length. However, we are aware that different domains, scenarios and AI models may influence the results we obtain. Another significant limitation comes from the generation of logical reasoning explanations. Although we employed well-established state-of-the-art methods, we acknowledge that different XAI techniques could be used as representatives of the reasoning styles, and this could produce varied outcomes. Further research may investigate the differences among different XAI techniques supporting the same reasoning style.

4 Implications on Engineering Interactive Systems

The research summarised in this paper shows how different characteristics of an AI-based decision support affect the user's performance in making a correct decision. The following is a list of criteria we can consider for making engineering decisions on the AI and explanation components grounded on the experiment results.

AI vs Task Perfomance. When introducing an AI-based component in an interactive system for suggesting a decision to the user, it is reasonable from an engineering point of view to consider usual performance metrics such as accuracy, precision, recall, etc. However, these measures tell how well the AI would perform if it takes the decision *alone*. From the results of the three experiments, it stands out that we cannot consider the characteristics of the AI model, the explanation technique and the user's uncertainty in isolation, but we must understand how they affect each other in determining the final decision. Unfortunately for us as engineers (but maybe fortunately for the overall decision), the goal of a decision support system is supporting the human in making a final correct decision. Introducing support that provides a suggestion may not be the best choice for establishing trust in the system, and this is why we usually add an explanation component trying to provide the user with means for interpreting the underlying motivation behind the AI's suggestion. But, as the data collected in the three experiments clearly show [3,4], the technique we use in the explanation component has an impact on the task performance, which depends not only on the AI model but also on the intrinsic characteristic of the domain. Therefore, the guidelines we discuss in this paper are tied to specific data types, domains, level of knowledge and other factors that influence the effectiveness of the decision support.

Instance Data Type. First of all, it is crucial to consider the data type we are working with. The intrinsic characteristic of the medium used for presenting the instance determines the different effects of the explanation style. For images and text, example-based (i.e., inductive style) explanations have a high persuasive power on users, leading them to agree with the AI no matter the correctness of its suggestion. However, this type of explanation does not work well when the data we want to describe embodies a complex representation, for example, when dealing with stock market charts (time series). Furthermore, explaining an instance by representing the features relevant to the AI's suggestion (abductive style) or general rules that lead the AI to its decision (deductive style) are effective ways to represent complex time-series data but are less effective for text and images.

User's Expertise. Knowing the target users' background is very useful to understand how effective the AI's suggestions and explanations may be. We noticed that domain knowledge increases the overconfidence in the user's decision, i.e., users rely only on their understanding of the presented data instance for making the prediction. While such a behaviour may be challenging to contrast, an engineering guideline may consider using a persuasive explanation style (e.g., inductive for images or text) if we expect the AI to perform better than the user in making the decision. This may induce the user to follow the AI more often.

While we collected evidence on the expertise aspect, it is worth mentioning that different user profile characteristics may have a role in the decision performance. For instance, Miller and Spatz [8] defined a taxonomy of metrics for representing the human characteristics affecting the effectiveness (i.e., the likelihood of mission completion) and efficiency (i.e., physical resources for mission completion) of a system. They relate human performance to the system performance, including, in the human-level taxonomy, dimensions such as expertise, fatigue, stress, vigilance, alertness and motivation. Each one of these dimensions may have an interaction with the selected explanation style.

AI Confidence. From a system engineering perspective, the most intriguing concept emerging from the study results is AI confidence. It possesses an important characteristic that makes it very useful in grounding engineering decisions: we can compute it on the instance we are considering. In brief, we can obtain the uncertainty by training the model using dropout levels, which is a common regularisation technique for avoiding overfitting [13]. Furthermore, we can exploit dropout during the evaluation phase for estimating the model uncertainty. This measure is computed on the current instance, similarly to the prediction. In contrast to other performance measures evaluating AI correctness, we can compute the AI confidence on a given instance without having access to the corresponding ground truth value. Therefore, estimating the AI confidence represents a beneficial feature for driving the *current* decision.

While the AI uncertainty effect on the different explanation styles was not consistent in all the domains we analysed, it always had a significant effect on the performance, agreement or both. So, even though it is difficult to provide guidelines in the general case, we suggest studying or considering the available results for the considered domain and data type to foster the desired user's behaviour under the current circumstances (e.g., increasing the agreement in case of a very precise AI model and high AI confidence, or relying more on the user's judgment in case of a low AI confidence etc.).

5 Conclusion and Future Work

This paper summarised the research we carried out on the effect of using explanations supporting different reasoning styles in three domains: stock market prediction, image classification and text sentiment recognition. We investigated the role of these explanations considering the users' uncertainty, AI correctness, and confidence.

We discuss different interesting results that can affect the engineering of systems including AI-based modules supporting users in making decisions. First of all, the reasoning style should be carefully selected according to the current task. There is suggestive evidence that the same reasoning style might be harmful for certain data types while it may increase the users' performance in others. Some of them are very powerful in convincing the users to follow the AI, while others may change their effectiveness according to the AI's confidence. Considering that the characteristics of the decision domain are known at design time and that the AI confidence may be obtained at runtime similarly to the prediction, we are positive that our research may contribute to the proper engineering of such systems.

In future work, we plan to investigate the effectiveness of different explanation styles in various domains and their correlation with the type of data presented. We will also explore the impact of other XAI techniques, such as counterfactual reasoning, within the same reasoning style. Additionally, we aim to examine the relationship between domain expertise, AI confidence, and task performance correctness. By addressing these questions, we aim to improve our understanding of explanation styles, XAI techniques, and their influence on decision-making processes and acceptance of AI suggestions.

Acknowledgements. This research has been carried out in collaboration with Hanna Hauptmann (Utrecht University) and Nava Tintarev (Maastricht University).

References

1. Asghar, N.: Yelp dataset challenge: review rating prediction (2016)
2. Buçinca, Z., Lin, P., Gajos, K.Z., Glassman, E.L.: Proxy tasks and subjective measures can be misleading in evaluating explainable AI systems. In: Proceedings of the 25th International Conference on Intelligent User Interfaces, March 2020. https://doi.org/10.1145/3377325.3377498
3. Cau, F.M., Hauptmann, H., Spano, L.D., Tintarev, N.: Effects of AI and logic-style explanations on users' decisions under different levels of uncertainty. ACM Trans. Interact. Intell. Syst. (2023). https://doi.org/10.1145/3588320, just Accepted

4. Cau, F.M., Hauptmann, H., Spano, L.D., Tintarev, N.: supporting high-uncertainty decisions through AI and logic-style explanations. In: Proceedings of the 28th International Conference on Intelligent User Interfaces, IUI 2023, pp. 251–263. Association for Computing Machinery, New York, NY, USA (2023). https://doi.org/10.1145/3581641.3584080
5. Hatwell, J., Gaber, M.M., Azad, R.M.A.: CHIRPS: explaining random forest classification. Artif. Intell. Rev. **53**(8), 5747–5788 (2020). https://doi.org/10.1007/s10462-020-09833-6
6. LeCun, Y., Cortes, C.: Mnist handwritten digit database (2010). http://yann.lecun.com/exdb/mnist/
7. Lundberg, S.M., Lee, S.I.: A unified approach to interpreting model predictions. In: Proceedings of the 31st International Conference on Neural Information Processing Systems, NIPS 2017, pp. 4768–4777. Curran Associates Inc., Red Hook, NY, USA (2017). https://doi.org/10.5555/3295222.3295230
8. Miller, M.E., Spatz, E.: A taxonomy of metrics for human representations in digital engineering. Systems Engineering **n/a**(n/a). https://doi.org/10.1002/sys.21722
9. Miller, T.: Explanation in artificial intelligence: insights from the social sciences. Artif. Intell. **267**, 1–38 (2019)
10. Ribeiro, M., Singh, S., Guestrin, C.: Why should i trust you?": Explaining the predictions of any classifier, pp. 97–101, February 2016. https://doi.org/10.18653/v1/N16-3020
11. Selvaraju, R.R., Cogswell, M., Das, A., Vedantam, R., Parikh, D., Batra, D.: Gradcam: visual explanations from deep networks via gradient-based localization. Int. J. Comput. Vis. **128**(2), 336–359 (2019). https://doi.org/10.1007/s11263-019-01228-7
12. Shrikumar, A., Greenside, P., Kundaje, A.: Learning important features through propagating activation differences, pp. 3145–3153 (2017)
13. Srivastava, N.: Improving neural networks with dropout. Univ. Toronto **182**(566), 7 (2013)
14. van der Waa, J., Nieuwburg, E., Cremers, A., Neerincx, M.: Evaluating xai: a comparison of rule-based and example-based explanations. Artif. Intell. **291**, 103404 (2021). https://doi.org/10.1016/j.artint.2020.103404, https://www.sciencedirect.com/science/article/pii/S0004370220301533
15. Vinyals, O., Toshev, A., Bengio, S., Erhan, D.: Show and tell: a neural image caption generator. 2015 IEEE Conference on Computer Vision and Pattern Recognition (CVPR), pp. 3156–3164 (2015)

Exploring AI-Enhanced Shared Control for an Assistive Robotic Arm

Max Pascher[1,2](✉), Kirill Kronhardt[1], Jan Freienstein[3], and Jens Gerken[1]

[1] TU Dortmund University, Dortmund, NW, Germany
max.pascher@udo.edu
[2] University of Duisburg-Essen, Essen, NW, Germany
[3] Westphalian University of Applied Sciences, Gelsenkirchen, NW, Germany

Abstract. Assistive technologies and in particular assistive robotic arms have the potential to enable people with motor impairments to live a self-determined life. More and more of these systems have become available for end users in recent years, such as the *Kinova Jaco* robotic arm. However, they mostly require complex manual control, which can overwhelm users. As a result, researchers have explored ways to let such robots act autonomously. However, at least for this specific group of users, such an approach has shown to be futile. Here, users want to stay in control to achieve a higher level of personal autonomy, to which an autonomous robot runs counter. In our research, we explore how Artificial Intelligence (AI) can be integrated into a shared control paradigm. In particular, we focus on the consequential requirements for the interface between human and robot and how we can keep humans in the loop while still significantly reducing the mental load and required motor skills.

Keywords: Assistive Technologies · Human-Robot-Interaction Mixed Reality · Shared Control · Visual Cues

1 Introduction and Motivation

When controlling an assistive robotic arm, one major challenge from a human-robot interface perspective is the mapping between available input controls and resulting robot movements. Assistive robotic arms capable of performing arbitrary pick-and-place tasks in 3D-space require at least seven Degrees-of-Freedom (DoFs): x-, y- and z-translation, roll, pitch, yaw, and opening/closing the robot's fingers (*cardinal* DoFs). Therefore, there is no direct mapping possible with most input devices. As a result, the user needs to constantly switch between modes, i.e., flip through several pre-defined mappings between input and output space. Research has shown that such mode switching requires a considerable amount of time and cognitive demand [2,5,12].

This research is supported by the *German Federal Ministry of Education and Research* (BMBF, FKZ: 16SV8565).

Much research in assistive robotics is concerned with autonomous robotic functions [4,10,18,30]. From that perspective, the mode switching dilemma may seem moot with robots gaining more and more autonomy through advanced Artifical Intelligence (AI) and thereby reducing the need for such direct control altogether. However, studies have shown that people with motor impairments may prefer manual control over autonomous execution, as they see the robot not so much as another agent but as a tool to regain self-determination (e.g., [16,28]).

To tackle this dilemma, different approaches along the continuum of shared control methods were proposed (e.g., [8,11,12,15,17,27,29,33]). The concept of shared control has great potential to design communication and control between humans and robots [1]. For assistive robotic arms, however, the various approaches address the issue on different levels, sometimes reducing the involvement of the user to simply indicate an object to be picked [33]. Other approaches directly addresses the mode switching issue, with Herlant et al. suggesting time-optimal mode switching along the cardinal DoFs [12] and – based on Dijkstra's algorithm – to predict when the robot should switch modes.

Recently, Goldau & Frese proposed an approach integrating a Convolutional Neural Network (CNN), which interprets live camera data from the gripper to constantly describe the probabilistic distribution of intended DoF robot motion and, accordingly, optimal mapping of DoFs [11]. In principle, the idea is that the user gets a suggestion not just for when and how to switch mode but going beyond cardinal DoFs suggestions, allowing more flexible and adaptive DoF combinations.

In our work, we explored the feasibility of such a CNN-based approach, identified empirical implications for shared control systems, and what kind of human-robot interaction design is feasible [17,24,27]. For this paper, we present the main challenges we identified and how our work has aimed to address these. In summary, these challenges are:

- **AI legibility:** Given an AI which is able to automate mode switches, the user needs to understand the behavior and actions of the AI. A system which changes input mappings without notification or explanation would be perceived as unpredictable.
- **AI user control:** Given an AI which is able to automate mode switches, the user must stay in the loop and have a final say in making the choices.
- **AI intervention:** Given an AI which is able to automate mode switches, the user needs to expect and prepare for AI mistakes. Therefore, they need to interfere with the AI and, at best, make it reconsider the mode switching.

2 Engineering Context: Simulation Environment

In our research, we found that – given the complexity of robotic arms in combination with the limitations of current AI technologies – a testbed environment that allows the integration of different control mechanisms and user interface components facilitates engineering.

To that end, we developed *AdaptiX* [24], a transitional XR framework for shared control applications in assistive robotics. *AdaptiX* resembles a real-world scenario, where an assistive robotic arm (here a *Kinova Jaco 2*) is used to facilitate pick-and-place tasks as we observed that they are part of many Activities of Daily Living (ADLs). The system combines a physical robot implementation with a 3D simulation environment. This approach, reminiscent of simulations used in industrial contexts [19,22,32], helps to address challenges associated with bulky, expensive, and complex assistive robotic arms. Researchers are empowered to streamline their Design and Development (D&D) process, reducing complexity and enhancing efficiency. The system's integrated Robot Operating System (ROS) interface enables seamless connectivity to a physical robotic arm, supporting bidirectional interactions and data exchange through a *DigitalTwin* and *PhysicalTwin* approach.

In addition to Cartesian robot control, the framework includes Adaptive DoF Mapping Controls (ADMC), an initial shared control approach that employs AI-generated suggestions, subject to user approval and control. Figure 1 provides an overview of the framework's architecture.

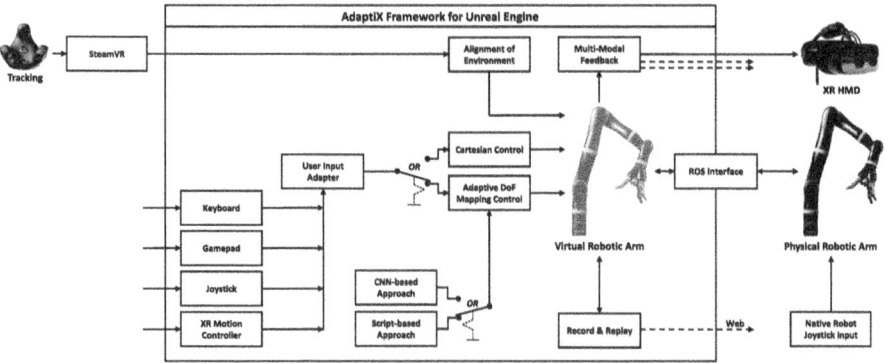

Fig. 1. Overview of *AdaptiX'* architecture, illustrating each component, their directional communication, and the crossover from and to the framework [24].

AdaptiX is built on the foundation of the game engine *Unreal Engine 4.27* [7]. This game engine is renowned for its advanced real-time 3D photorealistic visuals and immersive capabilities, making it an ideal choice for our framework. Furthermore, it offers a rich set of assets that can be readily used for future expansions. *Unreal Engine* is versatile and supports a variety of hardware configurations, allowing the framework to be deployed across different operating systems. It is compatible with a wide range of Virtual Reality (VR), Mixed Reality (MR), and Augmented Reality (AR) headsets, as well as gamepads and joysticks, making it suitable for development in both *C++* and *Blueprints*.

In the default scenario within *AdaptiX*, the focus is on a room that has been meticulously scanned using photogrammetry techniques. This room contains a table with an integrated virtual robotic arm, as depicted in Fig. 2. The simulation of the robotic arm has been optimized for operation via a VR motion controller, which features an analog stick, several functional buttons, and motion capture capabilities. An example of a compatible motion controller is the *Meta Quest 2* [20].

As most real-world scenarios will include pick-and-place operations [9,16,31,35], we designed a straightforward testbed scenario which requires to move a blue block to a red target area.

Fig. 2. Virtual environment consisting of (left to right): a virtual canvas, the motion controller, a table with a blue object and red target, and a *Kinova Jaco* with an arrow-based visualization [27].

We integrated the *Varjo XR-3* [34], a high-resolution XR Head-Mounted Displays (HMD), to create a seamless MR environment. By employing two *HTC VIVE* trackers [14], we synchronized the virtual and real worlds, ensuring that the operational spaces of the robots were perfectly aligned. This synchronization enables the adjustment of the MR level in multiple increments, as outlined in the *virtuality continuum* proposed by Milgram and Kishino [21]. A visual comparison between the user's perspective in the real world and the simulation is presented in Fig. 3.

The MR continuum comprises different levels. Level (**1**) serves as the study's baseline condition, offering no multi-modal feedback to the user. At level (**2**), the system mimics an AR visualization technique, which overlays the entire physical setup with basic cues. Especially level (**3**) and (**4**) enable customizing either the robot itself or the environment to extent/exchange the physical setup but still not loosing the context. In (**3**) users can interact with a totally new or customized robot while being in a familiar environment. World's distractions can be excluded in (**4**) while the original robot is presented. Level (**5**) provides a VR environment that is entirely customizable.

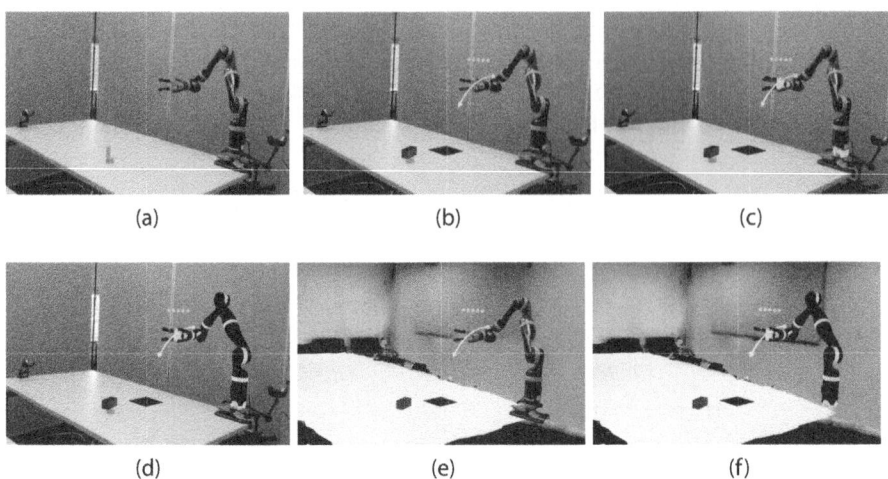

Fig. 3. MR continuum with (**a**) only the real robotic arm in real environment, (**b**) augmenting of directional cues in the real environment with the real robotic arm, (**c**) additional visualizing the gripper and base of the virtual robotic arm in the real environment, (**d**) visualizing the simulated robotic arm in the real environment, (**e**) visualizing the real robotic arm in the virtual environment, and (**f**) visualizing the simulated robotic arm in the virtual environment [24].

3 AI-Enhanced Shared Control

This section will provide more details about our previous research and how we addressed the three main challenges stated at the end of Sect. 1 – *AI legibility*, *AI user control*, and *AI intervention*. Therefore, we illustrate initial design concepts, work in progress, and prototypes that were evaluated in user studies.

3.1 AI Legibility

In the context of our research, achieving a level of AI legibility is mostly concerned with making it easier to understand how the AI would reassign the input mapping and/or change the movement direction of the robot. In our recent survey on such robot motion intent approaches [25], we found that, for communicating location information (such as a movement direction), head-mounted technology such as AR HMDs allow to represent the robot movement visually and have shown to provide a potential fruitful approach [6]. Although research has explored robot motion intent, there needs to be more insight into what works best in various situations and for different user types. Customizing the visualization and feedback modality is crucial, as there is no "one size fits all" solution [13].

We proposed different design concepts that fall into a spectrum with two extremes – indicative and explanatory [26]. **Indicative:** Focus on crucial information only, quick and easy solution, suitable for experienced robot users.

Explanatory: Movements are shown in great detail, high level of information, especially helpful for new users.

DoF-Indicator: LEDs attached to the robot's axis and joints – or mounted on a bar in front of it – communicate active and nonactive DoFs (see Fig. 4). Likely more suitable for experienced users because of indirect communication of movement, it communicates the DoF mapping and resulting movement abilities.

Fig. 4. DoF-Indicator: (a) LEDs directly attached at each robot's joint; (b) LEDs mounted on a bar in front of the robot referring to each joint (1–7) [26].

DoF-Combination-Indicator: Movement ability is communicated by a simplified representation of the robot only able to move two DoFs simultaneously, for example, rotating and extending the arm (see Fig. 5). The AR representation either overlays the real robot or can be displayed separately in the corner of the AR screen. This visualization decreases the robot's complexity.

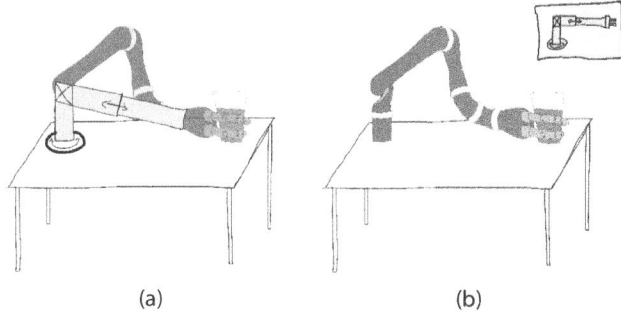

Fig. 5. DoF-Combination-Indicator: (a) as an AR overlay, supporting robot and visualization in line of sight; (b) as an icon in the screen's corner [26].

Gizmo Visualization: Arrows, planes and point clouds communicate the current movement ability of the robot (see Fig. 6). This allows for several different

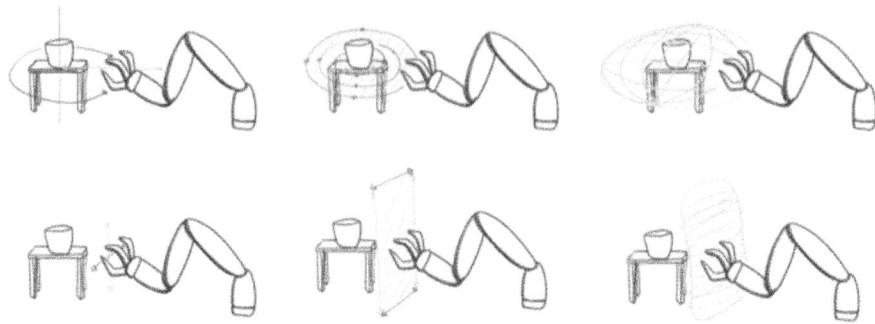

Fig. 6. Gizmo Visualization: **(left)** simple: straight and curved arrows; **(center)** planar: planes of movement; **(right)** cloud: 3D-cloud of possible positions [26].

design options. A arrow-based approach was already successfully evaluated in previous studies [17,27].

Demonstration: Current movement possibilities are demonstrated through either the actual robot or an AR ghost-representation. With both options a quick movement indicates the intended motion.

Visualization Approaches. Based on our initial concepts, we have explored augmenting the users' view with directional movement cues both in true – three-dimensional – AR (registered in 3D, [17,27]) as well as in 2D as symbolic representations on a data glass (for both refer to Fig. 7).

The former allows information-rich visualization and has shown in our studies to allow users to sufficiently anticipate a new suggested input mode mapping and corresponding movement direction [27].

The latter provides the advantage that the technology is already market ready and the devices are lightweight to carry and relatively easy to set up (compared to AR HMDs). However, without the ability to display directional visual cues registered in 3D space, the visual feedback is separated from the interaction space (robot) and may be more difficult to align with the current robot movement. In our work, we are currently exploring different visual forms, as can be seen in Fig. 7 c) and d).

In addition, we have been exploring other modalities which could either replace or complement visual feedback to increase the legibility of the AI. In [23], we explored different designs for vibrotactile feedback to communicate three-dimensional motion directions. We developed two conditions based on the *Cutaneous Rabbit* illusion and one based on *Apparent Tactile Motion* to communicate 2D direction. The gradient of the overall 3D direction was then encoded by the number of discrete vibration pulses, the vibration intensity, or a combination of both. Our study showed that three-dimensional directional cues could be communicated with a high success rate for both the 2D direction and gradient, but

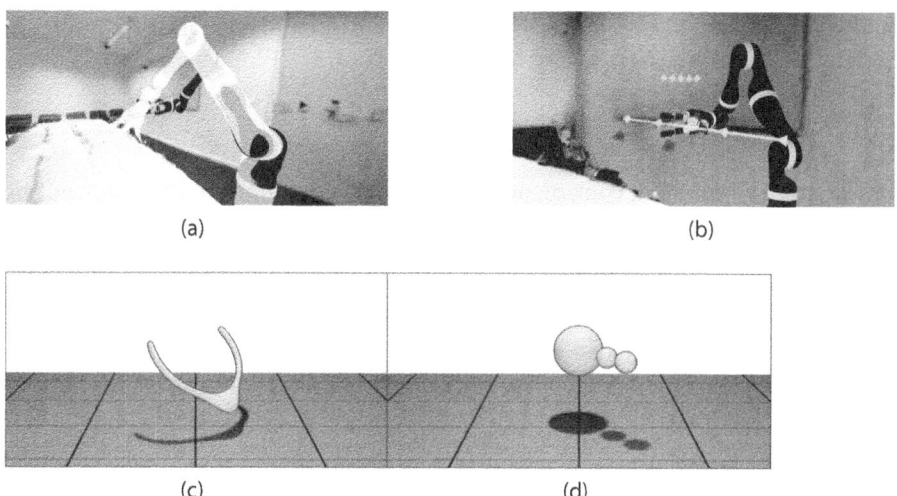

Fig. 7. Visualization examples for directional cues of an AI-supported robotic control in a VR 3D environment: (**a**) Ghost & (**b**) Arrows [24], and a real-world setting via data glasses, e.g., *Google Glass EE2*: (**c**) Ring & (**d**) Points.

may benefit from dual-encoding of the gradient information as well as individual customization of the specific implementation of the vibrotactile feedback patterns (see Fig. 8).

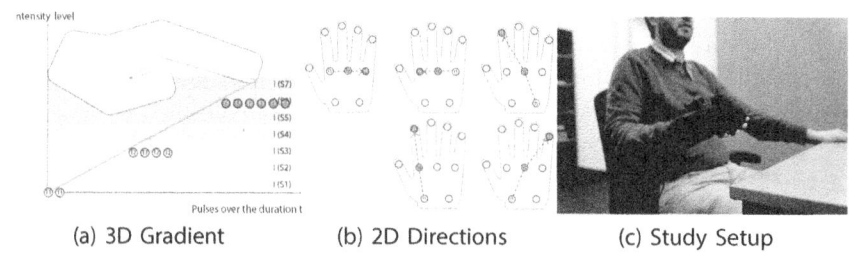

(a) 3D Gradient   (b) 2D Directions   (c) Study Setup

Fig. 8. Vibrotactile directional cues (**a**) 3D Gradient encoding with pulses and intensity mapping (**b**) 2D direction encoding across the hand (**c**) illustrates the study setup, with the arm resting on the armrest while the hand is in the air [23].

Empirical Implications. Through related work and our own research, we identified that – while the robot system is designed to act in the user's best interest – the user still needs to build trust, which requires transparency and legibility that they can comprehend. They should also be able to interfere with the robot's control if the robot makes a mistake or gives inappropriate suggestions for interaction. Communicating intent further requires having the user pay attention or guiding the attention of the user, requiring multi-modal stimuli, depending on the situation and the capabilities of the user.

3.2 AI User Control

For the user to remain in control, automatic mapping of input modes may not be desirable. Instead, we have explored different ways which allow users to stay in control but also benefit from the potential increase in efficient task completion. In their original approach, Goldau & Frese [11] asked users to wait for five seconds without interacting with the robot to trigger a new mapping.

As this has shown to cause some level of frustration, we explored ways for the user to directly request a new mapping [17] as well as integrating a continuous or threshold-based feedforward visualization of an updated mapping [27]. The latter has shown positive effects, as the user thereby can always compare the current movement and mapping with an update from the AI, but only switch to that when they feel that it changes the movement direction to achieve the task better.

In both studies, *Classic* – a non-adaptive control mode inspired by the *Kinova Jaco 2* standard joystick input – relies on mode switching to access and control all DoFs one after another and was used as a baseline condition. In comparison to *Classic*, our AI-based ADMC methods significantly reduced **(1)** the task completion time, **(2)** the average number of necessary mode switches, and **(3)** the perceived workload of the user.

Users may have diverse input device preferences and capabilities. This calls for the availability of multi-modal input options or the ability to choose between different input modalities [3]. To enable such user input, our simulation environment provides a standard control approach where pressing a keyboard button moves the end effector along cardinal DoFs (x, y, z, roll, pitch, yaw, opening and closing the gripper). Using further build-in functionalities, the designated keyboard input can easily be adjusted to other input devices like gamepads, joysticks, or customized assistive input appliances.

Empirical Implications. While the goal is to keep users in control, the complexity of both the robot interaction and the DoF limitations for available input devices can easily make the system very difficult to use. Therefore, in order to find the sweet spot for shared control, we propose to start with a rather minimized set of user interaction and increase that on demand and depending on the individual capabilities. Users should not be confused by too many interaction options or overly complex movements. While optimal ways of accomplishing a goal may require complex intervention from the robot, these interventions may be difficult for users to understand, and therefore trust. In addition, keep the DoF of input devices low as this maximizes the amount of assistive devices capable of controlling the robot.

Our previous research and related work show that pick-and-place tasks are ubiquitous and necessary to perform ADLs. It is, therefore, important that shared control is first implemented for these simple tasks before more complex sequences are examined. If users struggle to understand shared controls for pick-and-place tasks, we believe it is highly likely that more complex tasks may cause further frustration.

3.3 AI Intervention

While our approaches for AI user control allow some level of intervention, since the user can decide when to accept the updated input mapping as provided by the AI, it has some limitations. If the AI algorithm – as suggested by Goldau & Frese – is not able to provide a useful mapping, the user may become stuck with little flexibility to trigger the AI to update the mapping. We are currently exploring different approaches to tackle this issue for the specific shared control mechanism. One rather straightforward approach would be to allow the user to disable the AI and go back to manual mode switching of Cartesian DoFs. This of course may decrease the acceptance and perceived usefulness of the AI.

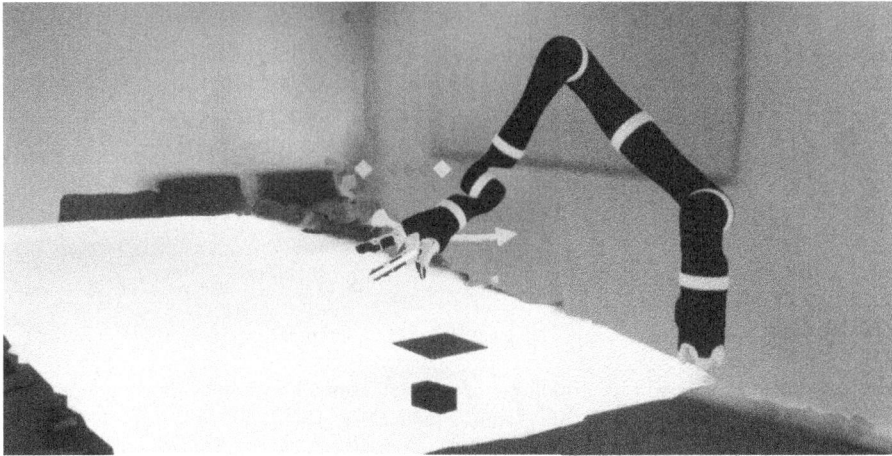

Fig. 9. The current DoF mapping (cyan arrow) does not allow to move to the blue object. By *changing the perspective* (green mode and arrow), the gripper is rotated in place to allow the CNN to suggest a new DoF mapping by an updated camera feed. (Color figure online)

A different way, which directly builds on the understanding of how the AI, in this case a CNN, operates, would be to find a way for the AI to *change perspective* – quite literally (see Fig. 9). If the robotic arm, or more specifically the gripper with the integrated or mounted camera is triggered to perform a small location repositioning, basically resulting in the robot looking around, the CNN will receive a new input which may result in a new and potentially better mapping.

Especially when the CNN is confronted with more than one choice (e.g., several objects in sight and vicinity of the gripper/camera) and has to choose one of them, based on the AI algorithm, rather than resulting in a draw. This kind of deadlock would only be resolved by a manual – Cartesian DoF – control. More suitable would be to either increasing AI's confidence of the chosen object by user input in the selected direction or decreasing by moving away, leading to a re-calculation and updated DoF mapping suggestion.

Empirical Implications. The aim is to keep the user in the loop so that they can intervene appropriately whenever the AI reaches its limits. However, this must strive for a balance that does not place sole decision dependency on the user to avoid access cognitive demand and temporal delays. Instead, establishing a four-eye principle with the AI functioning with implicit user consent until intervention is the most efficient approach to fulfilling the task's goal.

4 Conclusion

In this paper, we summarized our experiences for engineering AI-enhanced shared-control methods for assistive robotic arms. In particular, we identified three main challenges in *AI legibility*, *AI user control*, and *AI intervention*. Our work highlights the benefits and importance of sensible interaction design, which addresses these challenges and requires both a deep understanding of and interconnection with the AI technology. We also found that there is still much to be explored, in particular in the area of *AI intervention* approaches which go beyond circumventing the AI.

References

1. Abbink, D.A., et al.: A topology of shared control systems—finding common ground in diversity. IEEE Trans. Hum.-Mach. Syst. **48**(5), 509–525 (2018). https://doi.org/10.1109/thms.2018.2791570
2. Al-Halimi, R.K., Moussa, M.: Performing complex tasks by users with upper-extremity disabilities using a 6-DOF robotic arm: a study. IEEE Trans. Neural Syst. Rehabil. Eng. **25**(6), 686–693 (2017). https://doi.org/10.1109/TNSRE.2016.2603472
3. Arévalo-Arboleda, S., Pascher, M., Baumeister, A., Klein, B., Gerken, J.: Reflecting upon participatory design in human-robot collaboration for people with motor disabilities: challenges and lessons learned from three multiyear projects. In: The 14th PErvasive Technologies Related to Assistive Environments Conference - PETRA 2021. ACM (2021). https://doi.org/10.1145/3453892.3458044
4. Canal, G., Alenyà, G., Torras, C.: Personalization framework for adaptive robotic feeding assistance. In: Agah, A., Cabibihan, J.-J., Howard, A.M., Salichs, M.A., He, H. (eds.) ICSR 2016. LNCS (LNAI), vol. 9979, pp. 22–31. Springer, Cham (2016). https://doi.org/10.1007/978-3-319-47437-3_3

5. Chung, C.S., Wang, H., Cooper, R.A.: Functional assessment and performance evaluation for assistive robotic manipulators: literature review. J. Spinal Cord Med. **36**(4), 273–289 (2013). https://doi.org/10.1179/2045772313Y.0000000132
6. Cleaver, A., Tang, D.V., Chen, V., Short, E.S., Sinapov, J.: Dynamic path visualization for human-robot collaboration. In: Companion of the 2021 ACM/IEEE International Conference on Human-Robot Interaction, HRI 2021 Companion, pp. 339–343. Association for Computing Machinery, New York (2021). https://doi.org/10.1145/3434074.3447188
7. Epic Games: Unreal engine. Published online (2023). https://www.unrealengine.com. Version 4.27.2. Accessed 12 July 2024
8. Ezeh, C., Trautman, P., Devigne, L., Bureau, V., Babel, M., Carlson, T.: Probabilistic vs linear blending approaches to shared control for wheelchair driving. In: 2017 International Conference on Rehabilitation Robotics (ICORR), pp. 835–840 (2017). https://doi.org/10.1109/ICORR.2017.8009352
9. Fattal, C., Leynaert, V., Laffont, I., Baillet, A., Enjalbert, M., Leroux, C.: SAM, an assistive robotic device dedicated to helping persons with quadriplegia: usability study. Int. J. Soc. Robot. **11**(1), 89–103 (2019). https://doi.org/10.1007/s12369-018-0482-7
10. Gallenberger, D., Bhattacharjee, T., Kim, Y., Srinivasa, S.S.: Transfer depends on acquisition: analyzing manipulation strategies for robotic feeding. In: 2019 14th ACM/IEEE International Conference on Human-Robot Interaction (HRI), pp. 267–276 (2019). https://doi.org/10.1109/HRI.2019.8673309
11. Goldau, F.F., Frese, U.: Learning to map degrees of freedom for assistive user control: towards an adaptive DOF-mapping control for assistive robots. In: The 14th PErvasive Technologies Related to Assistive Environments Conference, PETRA 2021, pp. 132–139. Association for Computing Machinery, New York (2021). https://doi.org/10.1145/3453892.3453895
12. Herlant, L.V., Holladay, R.M., Srinivasa, S.S.: Assistive teleoperation of robot arms via automatic time-optimal mode switching. In: 2016 11th ACM/IEEE International Conference on Human-Robot Interaction (HRI), pp. 35–42 (2016). https://doi.org/10.1109/HRI.2016.7451731
13. Holloway, C.: Disability interaction (DIX): a manifesto. Interactions **26**(2), 44–49 (2019). https://doi.org/10.1145/3310322
14. HTC: Vive tracker 3.0. Published online (2023). https://www.vive.com/de/accessory/tracker3/. Accessed 12 July 2024
15. Jain, S., Farshchiansadegh, A., Broad, A., Abdollahi, F., Mussa-Ivaldi, F., Argall, B.: Assistive robotic manipulation through shared autonomy and a body-machine interface. In: 2015 IEEE International Conference on Rehabilitation Robotics (ICORR), pp. 526–531 (2015). https://doi.org/10.1109/ICORR.2015.7281253
16. Kim, D.J., et al.: How autonomy impacts performance and satisfaction: results from a study with spinal cord injured subjects using an assistive robot. IEEE Trans. Syst. Man Cybern. - Part A: Syst. Hum. **42**(1), 2–14 (2012). https://doi.org/10.1109/TSMCA.2011.2159589
17. Kronhardt, K., Rübner, S., Pascher, M., Goldau, F., Frese, U., Gerken, J.: Adapt or perish? Exploring the effectiveness of adaptive dof control interaction methods for assistive robot arms. Technologies **10**(1) (2022). https://doi.org/10.3390/technologies10010030
18. Lauretti, C., Cordella, F., Guglielmelli, E., Zollo, L.: Learning by demonstration for planning activities of daily living in rehabilitation and assistive robotics. IEEE Rob. Autom. Lett. **2**(3), 1375–1382 (2017). https://doi.org/10.1109/LRA.2017.2669369

19. Matsas, E., Vosniakos, G.C.: Design of a virtual reality training system for human-robot collaboration in manufacturing tasks. Int. J. Interact. Design Manuf. (IJIDeM) **11**(2), 139–153 (2015). https://doi.org/10.1007/s12008-015-0259-2
20. Meta: Meta quest 2. Published online (2023). https://www.meta.com/de/quest/products/quest-2/. Accessed 12 July 2024
21. Milgram, P., Kishino, F.: A taxonomy of mixed reality visual displays. IEICE Trans. Inf. Syst. **77**(12), 1321–1329 (1994)
22. Müller, S.L., Schröder, S., Jeschke, S., Richert, A.: Design of a robotic workmate. In: Duffy, V.G. (ed.) DHM 2017. LNCS, vol. 10286, pp. 447–456. Springer, Cham (2017). https://doi.org/10.1007/978-3-319-58463-8_37
23. Pascher, M., Franzen, T., Kronhardt, K., Gruenefeld, U., Schneegass, S., Gerken, J.: HaptiX: vibrotactile haptic feedback for communication of 3D directional cues. In: Extended Abstracts of the 2023 CHI Conference on Human Factors in Computing Systems, CHI EA 2023. Association for Computing Machinery, New York (2023). https://doi.org/10.1145/3544549.3585601
24. Pascher, M., Goldau, F.F., Kronhardt, K., Frese, U., Gerken, J.: AdaptiX – a transitional XR framework for development and evaluation of shared control applications in assistive robotics. Proc. ACM Hum.-Comput. Interact. **8**(EICS) (2024). https://arxiv.org/abs/2310.15887
25. Pascher, M., Gruenefeld, U., Schneegass, S., Gerken, J.: How to communicate robot motion intent: a scoping review. In: Proceedings of the 2023 CHI Conference on Human Factors in Computing Systems, CHI 2023. Association for Computing Machinery, New York (2023). https://doi.org/10.1145/3544548.3580857
26. Pascher, M., Kronhardt, K., Franzen, T., Gerken, J.: Adaptive DoF: concepts to visualize AI-generated movements in human-robot collaboration. In: Proceedings of the 2022 International Conference on Advanced Visual Interfaces, AVI 2022. Association for Computing Machinery, New York (2022). https://doi.org/10.1145/3531073.3534479
27. Pascher, M., Kronhardt, K., Goldau, F.F., Frese, U., Gerken, J.: In Time and space: towards usable adaptive control for assistive robotic arms. In: 2023 32nd IEEE International Conference on Robot and Human Interactive Communication (ROMAN), pp. 2300–2307. IEEE (2023). https://doi.org/10.1109/RO-MAN57019.2023.10309381
28. Pollak, A., Paliga, M., Pulopulos, M.M., Kozusznik, B., Kozusznik, M.W.: Stress in manual and autonomous modes of collaboration with a cobot. Comput. Hum. Behav. **112**, 106469 (2020). https://doi.org/10.1016/j.chb.2020.106469
29. Quere, G., et al.: Shared control templates for assistive robotics. In: 2020 IEEE International Conference on Robotics and Automation (ICRA), pp. 1956–1962 (2020). https://doi.org/10.1109/ICRA40945.2020.9197041
30. Rakhimkul, S., Kim, A., Pazylbekov, A., Shintemirov, A.: Autonomous object detection and grasping using deep learning for design of an intelligent assistive robot manipulation system. In: 2019 IEEE International Conference on Systems, Man and Cybernetics (SMC), pp. 3962–3968 (2019). https://doi.org/10.1109/SMC.2019.8914465
31. Shafti, A., Orlov, P., Faisal, A.A.: Gaze-based, context-aware robotic system for assisted reaching and grasping. In: 2019 International Conference on Robotics and Automation (ICRA), pp. 863–869 (2019). https://doi.org/10.1109/ICRA.2019.8793804
32. Tidoni, E., et al.: Local and remote cooperation with virtual and robotic agents: a P300 BCI study in healthy and people living with spinal cord injury. IEEE

Trans. Neural Syst. Rehabil. Eng. **25**(9), 1622–1632 (2017). https://doi.org/10.1109/tnsre.2016.2626391
33. Tsui, K.M., Kim, D.J., Behal, A., Kontak, D., Yanco, H.A.: "I want that": human-in-the-loop control of a wheelchair-mounted robotic arm. Appl. Bion. Biomech. **8**(1), 127–147 (2011)
34. Varjo: Varjo XR-3 (2023). https://varjo.com/products/xr-3/. Accessed 12 July 2024
35. Wang, D., et al.: Towards assistive robotic pick and place in open world environments. In: Asfour, T., Yoshida, E., Park, J., Christensen, H., Khatib, O. (eds.) ISRR 2019. Springer Proceedings in Advanced Robotics, vol. 20, pp. 360–375. Springer, Cham (2022). https://doi.org/10.1007/978-3-030-95459-8_22

Hidden Figures: Architectural Challenges to Expose Parameters Lost in Code

Alan Dix[1,2]

[1] Computational Foundry, Swansea University, Swansea, Wales
alan@hcibook.com
[2] Cardiff Metropolitan University, Cardiff, Wales
https://alandix.com/academic/papers/EISEAIT2023-hidden/

Abstract. Many critical user interaction design decisions are made in the heat of detailed development. These include simple parameter choices or more complex weightings in intelligent algorithms. Many would be appropriate for expert design review, user-preference choices or optimisation by machine learning, but they are buried deep in the code. Although the developer may realise this potential, the location of the decision is far removed in the code from where user feedback occurs, data can be collected and machine learning could be applied. This position paper describes several case studies and uses them to frame an architectural challenge for tools and infrastructure to uncover these hidden variables to make them available for machine learning and user inspection.

Keywords: Intelligent interfaces · machine learning · user interface architecture

1 Motivation

Many critical user interaction design decisions are not made by user experience designers informed by extensive user research, but rather by a developer working to a deadline needing to make a choice in order to progress. Sometimes these are simple parameters such as the time-delay and pixel-precision for a double click or the default choice in a menu. Some are more complex such as sets of weightings within intelligent algorithms. Many would be appropriate for expert design review, user-preference choices or optimisation by machine learning. Some are identified and maybe tuned using techniques such as A–B testing. However, often these parameters, values and decisions are buried deep in the code.

Consider the following code fragment:

```
ON mouseup
   IF ABS( mousedown_x - current_x ) <= 2
           AND ABS( mousedown_y - current_y ) <= 2
   THEN TRIGGER click
```

This is configured to allow a maximum 2 pixel drift between mouse-down and mouse up for it to be considered a click rather than a move/drag. There will be similar constants for the number of milliseconds between two clicks for them to be regarded as a double click, or in a touch-based interface for press vs hold.

It is often said that in code the only fixed numeric values should be 0 and 1, any other parameters should be in some sort of constant declaration or configuration. Equally where there is an enumeration, any use of a literal enumeration value suggests a configuration setting, and a good coder might do this:

```
CONSTANT   click_pixel_precision = 2
CONSTANT   double_click_delay    = 500
CONSTANT   default_font          = "Times Roman"
```

Imagine if these fixed values could be semi-automatically exposed in a user-preferences dialog and/or made available for machine learning.

In fact, many systems do offer the ability to tune the double-click period in user preferences, whereas the author has never encountered one that allows tuning of the pixel threshold for a click. Indeed, the latter is often zero, no movement allowed, which is particularly problematic for older or younger users, and clearly not being made available to automated tuning either.

Whether these parameters are escalated to user preferences or automated optimisation depends partly on the developer recognising their importance, but also the difficulty of achieving this in most architectures.

In the case of user preferences, the path is slightly easier. The relevant configuration parameters can be made into variables in some form of central context or configuration object and set in a preferences dialog. Often this is easier to say than to achieve as the ownership both organisationally and in terms of code encapsulation may get in the way.

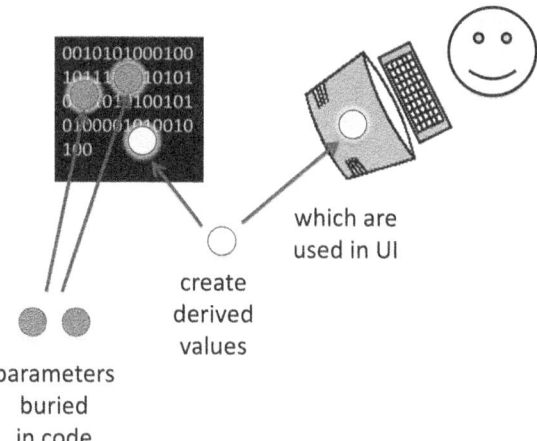

Fig. 1. Hidden values that are potentially user or AI modifiable, but buried deep in the code

For automated optimisation using artificial intelligence, the path is far harder as illustrated in Fig. 1. The parameters that could be modifiable by machine learning are buried deep in back-end code that computes some form of derived values, for example creating a score for ordering alternative options. These derived values are then used in the front-end user interface where it may be possible to obtain feedback on their effectiveness, for example, the actual option chosen. However, the point of computation and point of use are distant in the code.

In principle it is possible to add bespoke instrumentation at the point of use, collect and store this and then pass it into some form of continual or periodic batch learning algorithm the results of which can be made available for future execution of the code. However, every infrastructure, notation, toolkit and architectural paradigm has a 'grain', just like the grain of wood in carpentry [9]. It may be possible to work across the grain, but far more likely that developers will do the easy thing.

It has long been understood that the paths of connection in a user interface often cut across the module and encapsulation boundaries of functionally-defined architectures; hence the emergence and importance of frameworks such as MVC [12] or PAC [2]. More recently the availability of toolkits such as React (https://react.dev/) and Angular (https://angular.io/) has meant that real-time interactive editing of back-end data has become common, despite the inherent complexity, as this is managed within the frameworks.

Similarly, if we want good AI in UX, we need the right frameworks to make this possible.

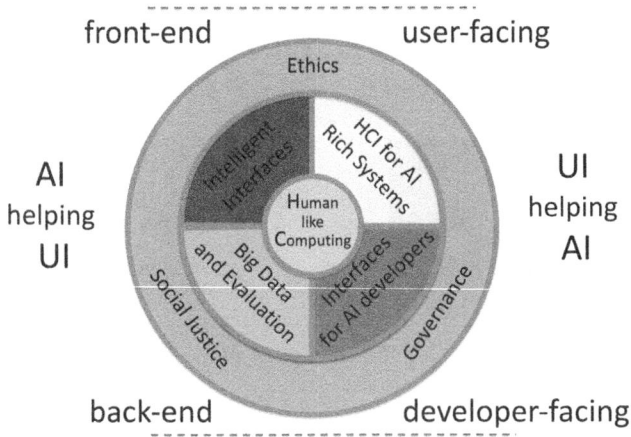

Fig. 2. The AI for HCI quadrants.

In the rest of this paper, we will look at three case studies of systems that are to some extent 'intelligent', but within each lie buried parameters and values, which could be amenable to machine learning to tune and improve the user experience. In terms of the quadrants of AI for HCI (Fig. 2), this work lies towards to the bottom right, that is HCI applied to help developers of AI systems, but where the ultimate goal of the AI development is in the upper quadrant, short and long-term interactions incorporating AI.

This paper presents an open problem for the community, with hints of a first direction, but no solution. The author hopes that this will inspire other researchers to identify similar patterns in their own work and maybe eventually work towards systematic and generic architectural paradigms for AI-rich interactions.

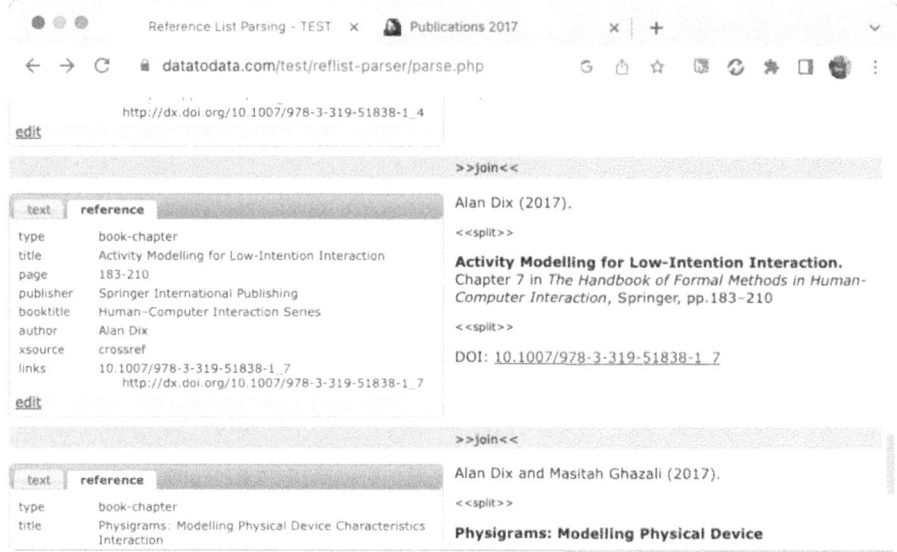

Fig. 3. Reference list parser: right-hand side is raw text; left-hand side parsed references.

2 Case Studies

These issues are not theoretical. Here are three case studies that the author has been directly involved in where they have arisen in practice.

2.1 Reference List Parsing

Figure. 3 shows a prototype interactive reference list parser that was produced some years ago, when the author was working for Talis, which creates reading list software for higher education. The parser is given a complete reference list from a paper or reading list for a course and makes initial best guesses at which lines to combine to make a single reference. Once the units have been decided, the reference is passed to the Crossref REST API [3] to produce structured reference data for the resulting blocks. Crossref returns highly accurate references, but avoids parsing as it effectively uses free-text search over structured bibliographic data [4].

The reference parser is not, however, purely automatic, but in fact employs synergistic interaction with the user. The user can fine-tune chosen blocks using the 'split' and 'join' separators, when this happens the automated part of the system re-checks the amended reference units with Crossref. In addition, while the outputs of Crossref are of high quality, the user can edit the returned references if they wish or fill in the details of any that Crossref has not found.

When the system was first created Crossref included few books and non-journal articles, and so other services were used to supplement it. A separate book search was implemented using Open Library data (https://openlibrary.org/) and Apache Solr free text index [14]. In addition, Freecite API (now retired) was used to parse the raw text references [1].

```
35     var $params = array(
36         'single_numeric_title_match_penalty' => 0.1,
37         'numeric_title_match_penalty' => 0.3,
38         'all_numeric_title_match_penalty' => 0.4,
39         'words_used_weight'=>0.2,
40         'combine_field_ct_weight'=>0.05,
41         'combine_all_fields_ct_weight'=>0.05,
```

```
117    var reflist_options = { bias: { text: 1, freecite: 1.5, crossref: 2 }, textscore: 100 };
```

Fig. 4. Reference list parser parameters: server side above (fragment), client side below

There could be results from more than one service, and some services, including Crossref, may return several responses, so the responses need to be ranked. To do this each result was assigned a relevance score. A number of metrics were used including whether the first author was in the free text reference, whether the words of the text fields in the structured reference included most of the words in the text, etc. Some metrics had internal parameters and there were also preferences between the various sources (see Fig. 4). However these were always 'best guesses' and never tuned even though the user's choices to override system selections could in principle have been used as feedback for machine learning.

2.2 Matching Historic Records

In the InConcert project a few years ago the author was working with musicologists looking at London concerts from the middle of the 18th Century to the beginning of the 20th Century. For more recent events concert programmes are

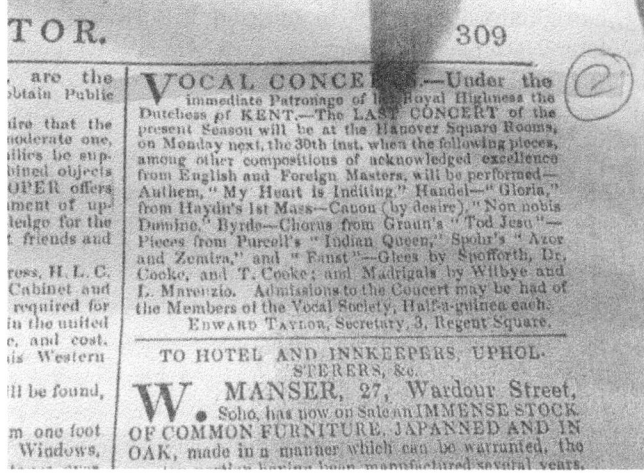

Fig. 5. Concert notice in 19th Century newspaper.

often extant, but for the 19th Century the main source of information is concert notices in newspapers, as seen in Fig. 5.

The data had been hand transcribed from the notices in a semi-structured form, but needed substantial interpretative work in order to make it usable for scholarly analysis. This included matching named entities (people and venues) within a dataset and between different datasets and also matching of different concert notices that refer to the same actual concert. Both kinds of matching need to be fuzzy. Considering named entities, personal names may use initials or full names and have different honorifics; similarly places may have variant spellings and descriptions. Events are more complex as there may be some or all of location, date, time, concert title, and there may be variations in many including of course that the venues may only fuzzily match.

Automatic matches were made and ranked for the musicologists to hand-check. The aim was to get the automatic matching as good as possible, but also not to miss potential matches [6,8]. Just as in the case of the reference list parser, there were various hand-crafted individual features, such as matched dates and levels of partially matched concert titles, and many parameters and weights used to combine these features into a single score. Again these parameters and weights were 'best guesses' and never tuned even though in the next stage the musicologists would accept or reject matches which would, in principle, have given good feedback for machine learning of the parameters.

2.3 Data Detectors

The final example is from Snip!t, an experimental web service that was somewhere between a web scrapbook and extended bookmark service. It allowed segments of web pages to be snipped to save for later, but retaining their provenance link to the original web page [5]. In addition, it borrowed techniques from onCue, a dot-com era intelligent internet interface, which used data detectors (a simple form of intelligent matching) to find semantic information in unstructured text [7].

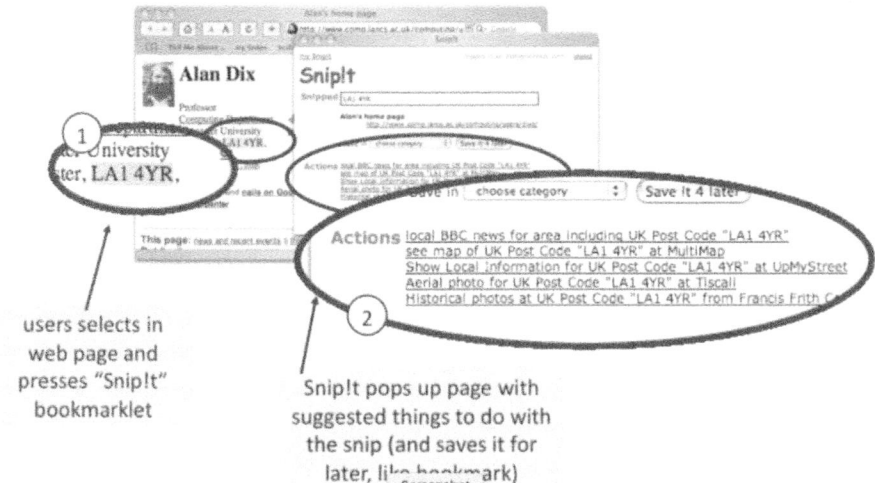

Fig. 6. Snip!t – how it works.

```
<beforeafterrecogniser>
  <name>name1Recogniser</name>
  <title>Person Name recogniser</title>
  <keyed>
    <keys>Female First Name, Male First Name</keys>
  </keyed>
  <pattern>
    <pre_context>\W+</pre_context>
    <before>($RE_HONOURIFICS\s*)?</before>
    <key>$RE_CAPSNAME</key>
    <after>\s*$RE_MIDNAMES\s*$RE_LASTNAME</after>
    <post_context>\W+</post_context>
  </pattern>
  <match>
    <type>name</type>
    <description>Person name $$</description>
  </match>
</beforeafterrecogr
```

Fig. 7. Data detectors in Snip!t: keyed name recogniser.

Figure 6 shows Snip!it in action. A post code on a web page has been 'snipped', the system recognises this and then suggests potential things to do with a postcode. The postcode recogniser is a simple regular expression:

```
([A-Za-z][A-Za-z0-9]{1,3})[ \t]{1,6}([0-9][A-Za-z]{2})
```

However, other recognisers are more complicated. Figure 7 is a proper name recogniser, which uses initial data-lookup of personal names derived from census records in order to trigger the recogniser, which then combines this with regular expressions.

The matching is a simple Yes/No, but really some matches are better than others, for example a name from the data source is more likely to be a real name than something that simply matches the pattern of an initial capital. However, the complexity of choosing and then tuning these meant it was never done.

3 Engineering for Hidden Variables – Architectural Challenges

As we have seen there are often rich opportunities for machine learning to optimise parameters of user interfaces. However, even when the developer is aware of this potential, the location of the decision is far removed in the code from where user feedback would occur. Code would have to be substantially reworked in order to enable the automation flows. If user interfaces are to make full use of AI potential, we need architectures and frameworks to enable this.

The examples, we have seen have a relatively similar structure. At the point the relevant parameters are noticed, they need to be marked so that they can be extracted by a suitable support framework. Once extracted, a UX designer can choose to present them in user preferences and/or mark them for automated learning. In the latter case it is also important that the uses of the relevant parameters are recorded in the code where they feed into dependent values, and also where these dependent values are presented to the user; this would enable user behaviour to be re-connected to the parameterisation for creating learning sets. The final stages of machine learning and then using this to update configuration are relatively straightforward.

One way to achieve this is through some form of globally accessible hidden value framework object/module. This could provide a mechanism at point of calculation that takes a list of parameters used and a function that does the actual calculation:

```
HV_Framework.calcDerived(
    names_of_params_used,
    calculation_function
)
```

The framework would extract the parameters, pass them to the actual calculation functions, and then store the parameters used and derived value in some form of log or database. The framework would then return a wrapped value that includes both the actual computed value and also an identifier or reference to the stored information.

When the wrapped value is needed it can be unwrapped for use in the user interface, and the framework reference linked to the relevant UI widget. This then means that feedback based on implicit or explicit user interaction can be provided by the application code back to the framework.

```
HV_Framework.feedback(
    wrapped_value,
    feedback    // e.g. 'smaller'/'larger'
)
```

The framework can then use the reference in the wrapped value to associate the feedback with the original stored parameters ready for machine learning.

This would require the developer to identify the relevant code locations and explicitly use the framework. It would also add some friction to the coding as wrapped values need to be unwrapped when used. However, this is easier than managing the complete process in a bespoke manner.

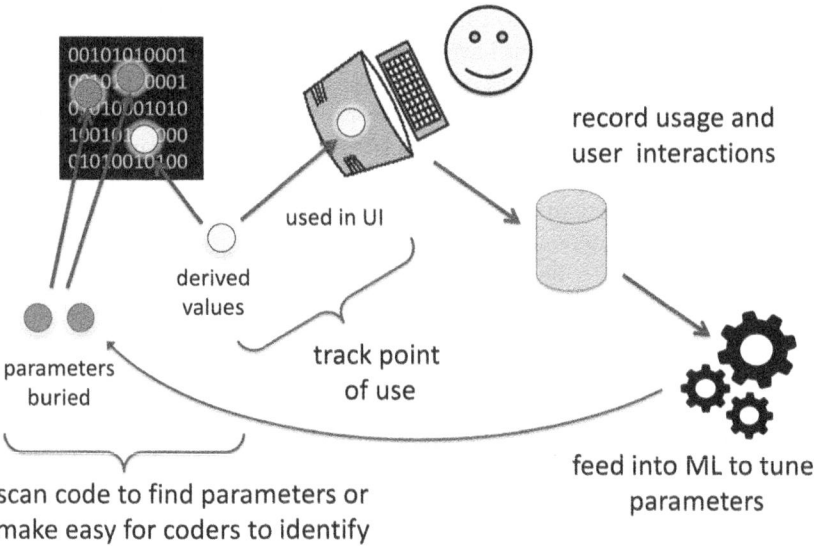

Fig. 8. Exposing hidden values

An alternative, would be to use some form of structured commenting so that a code-analysis tool can identify the points of production and of use and then insert automatically generated code (such as the above) to collect the data needed for machine learning, as illustrated in Fig. 8. This process is similar to the matching–weaving model in aspect-oriented programming [11]; indeed it maybe that existing AOP tools, such as AspectJ, could be used for this, although these have never become widely used. In addition, there are various well-used commercial frameworks that employ some form of source code annotations. Both Java annotations and .NET attributes provide customisable extensions using the same syntax as built-in compiler annotations; these can be processed at source level, but are also passed through to compiled code so that they can be interrogated at run time [10,13]. IDEs such as Eclipse support these language mechanisms as well as comment based annotations meaning that it is possible to create plugins to process annotations as part of a normal toolchain (e.g. [15]).

In both alternatives, there may be some additional analysis needed at the point of initial computation, for example to calculate gradients for each parameter used, either through symbolic analysis of the calculation function, or blackbox sensitivity calculations. The former would typically only be possible with code-analysis techniques, and then only for simpler calculations. The latter is more generally applicable, but would typically incur some additional computational overhead at run time.

4 Summary

We have seen several examples of the way parameters that could be tuned using machine learning are buried deeply in code and thus become frozen as the developer's 'best guess'. These all followed the rough pattern that we originally considered in Fig. 1. We have then discussed two engineering options for how this could be achieved either through a coding framework or automated analysis of structured code.

Of course this is not the only pattern of AI use in user interaction, so frameworks have to open and flexible enough for a wider variety of patterns. It may even be that AI techniques can be used as part of the analysis of existing code bases in order to uncover hidden UX variables for tuning.

Tho goal of this paper has been to highlight the challenges and potential of hidden variables for automated tuning in user interfaces and more general human interactions. It suggests some potential solution strategies, but the main take-away is that this is an area with great potential for research that is both theoretically interesting and of practical benefit.

References

1. Brown University: FreeCite – open source citation parser (2008). https://tinyurl.com/freecite. Accessed 29 May 2023
2. Coutaz, J.: PAC, an object oriented model for dialog design. In: Human–Computer Interaction–INTERACT 1987, pp. 431–436. Elsevier (1987)
3. Crossref: REST API (2023). https://www.crossref.org/documentation/retrieve-metadata/rest-api/. Accessed 10 Dec 2023
4. Crossref Labs: Resolving citations (we don't need no Stinkin' parser) (2017). Accessed 29 May 2023

5. Dix, A.: Context and action in search interfaces. In: Ceri, S., Brambilla, M. (eds.) Search Computing. LNCS, vol. 6585, pp. 35–45. Springer, Heidelberg (2011). https://doi.org/10.1007/978-3-642-19668-3_4
6. Dix, A.: Designing user interactions with AI: servant, master or symbiosis. The AI Summit London (2021). https://www.alandix.com/academic/talks/AI-Summit-2021-UI-with-AI/
7. Dix, A., Beale, R., Wood, A.: Architectures to make simple visualisations using simple systems. In: Proceedings of the Working Conference on Advanced Visual Interfaces, AVI 2000, pp. 51–60. ACM (2000). https://doi.org/10.1145/345513.345250
8. Dix, A., Cowgill, R., Bashford, C., McVeigh, S., Ridgewell, R.: Spreadsheets as user interfaces. In: Proceedings of the International Working Conference on Advanced Visual Interfaces, AVI 2016, pp. 192–195. ACM (2016)
9. Dix, A., Finlay, J., Abowd, G.D., Beale, R.: Design focus: going with the grain. In: Human-Computer Interaction, p. 301. Pearson Education (2003)
10. Hunter, J.: Making the most of Java's metadata, part 2: custom annotations. Oracle Technical Article (2023). https://www.oracle.com/technical-resources/articles/hunter-meta1.html. Accessed 10 Dec 2023
11. Kiczales, G., et al.: Aspect-oriented programming. In: Akşit, M., Matsuoka, S. (eds.) ECOOP 1997. LNCS, vol. 1241, pp. 220–242. Springer, Heidelberg (1997). https://doi.org/10.1007/BFb0053381
12. Krasner, G.E., Pope, S.T., et al.: A description of the model-view-controller user interface paradigm in the Smalltalk-80 system. J. Object Orient. Program. **1**(3), 26–49 (1988)
13. Microsoft: Extend metadata using attributes (2021). https://learn.microsoft.com/en-us/dotnet/standard/attributes/. Accessed 10 Dec 2023
14. Smiley, D., Pugh, E., Parisa, K., Mitchell, M.: Apache Solr Enterprise Search Server. Packt Publishing Ltd. (2015)
15. Styger, E.: To-do lists with Eclipse tasks view (2016). https://mcuoneclipse.com/2016/05/24/to-do-lists-with-eclipse-tasks-view/. Accessed 10 Dec 2023

Not What I was Trained for – Out-of-Distribution-Tests for Interactive AIs

Benedikt Severin(✉), Ole Werger, and Marc Hesenius

Institute for Software Engineering, University of Duisburg-Essen, Schützenbahn 70, 45127 Essen, Germany
{benedikt.severin,ole.werger,marc.hesenius}@uni-due.de

Abstract. Recent advances in AI demonstrate its capacities to not only automate more and more everyday tasks, but also make direct Human-AI-Interaction possible. When using AI models in production, we might face situations where the AI is confronted with input data that is very different from the data it was trained on. Such situations are called Out-of-Distribution situations and can result in misleading AI inferences. We argue that identification, handling, and prevention of Out-of-Distribution situations is key for creating production-ready interactive AI components. In this paper, we test the robustness of state-of-the-art AI/ML approaches in Out-of-Distribution situations and propose a research agenda to gather a deeper understanding of how to identify, handle, and prevent such situations in interactive applications.

Keywords: Human-Computer Interaction · Software Engineering · Artificial Intelligence · Machine Learning · Testing · Test Automation · Quality Assurance · Out-of-Distribution · Concept Shift

1 Introduction

Recent developments in generative Artificial Intelligence (AI), such as Large Language Model (LLMs) or text-to-image generators, offer great potential for new applications in Human-Computer Interaction (HCI), ranging from prompt-based generation of Graphical User Interfaces (GUIs) [7,35] to natural-language dialogs. The boundaries between dedicated AI solutions and classic software are fading and users might not even realize that they are interacting with an AI. On the one hand, AI components might be used as consumer-facing interfaces that accept, e.g., commands in (written or spoken) natural language and respond with a generated response. Users will likely be aware that they are interacting with an AI component. On the other hand, AI components might also be embedded software solutions for very specific problems like grammar auto correction in text processors, voice pitch correction for fast-forward voice messages in mobile messengers, or gesture recognizers. From an HCI perspective, AI components are employed to enhance the User Experience (UX) and integrate a more natural interaction with the machine. Ensuring the reliability of AI models is, however, critical for an acceptable UX.

Creating reliable AI applications requires high quality and diverse training data as well as dedicated Software Engineering (SE) methods (see, e.g., Hesenius et al. [23]). As most AI applications use Machine Learning (ML) techniques, their probabilistic nature introduces uncertainty into the application [22], but it is also uncertain with what data the AI will be confronted with after deployment. In simple terms, while developers may be able to ensure high data quality during development, they cannot control the data used for inference at runtime. Runtime data, however, is equally important to model performance and acceptance as the data used for training. For example, AI/ML is often used for gesture recognition [36], but the performance varies between user groups: recognizers are often trained with samples from healthy adults and the recognition performance decreases for younger children [2], because their lack of fine motor skills results in greater shape variability [43]. If a multi-touch application using an AI model for gesture recognition is operated by a child, the model would encounter input data that differs from the original training data in terms of its distribution. Such situations are thus called *Out-of-Distribution (OOD)* situations: during inference (after the model has been deployed), the model encounters valid yet somehow different data, resulting in decreased performance that impacts UX.

Such situations are well-known in SE: We refer to an application's ability to cope with faulty or unexpected input data as its *robustness*. A common test is flooding the application with random input data and inspecting it for potential crashes and unexpected behavior (also called *monkey testing*). Such an approach is feasible if no control over input is needed, but, for example, gesture-based interfaces require a more strategic approach as random input data cannot yield meaningful results [20]. We argue that testing AI/ML applications also requires control over testing data due to its influence on the UX.

OOD situations are a symptom from constraining models to their insufficiently diverse training data. Large pre-trained models are more stable in this regard compared with specialized models [16] due to their training on larger quantities of diverse data, but such data will only be available to a few companies. Use cases that handle sensitive information, such as medical data, might also have privacy constraints forbidding cloud-hosted large pre-trained models. For all these AI/ML approaches, companies must introduce dedicated methods into their development to hunt down possible OOD situations. Developers require methods to systematically check for the impact that OOD situations can have on their application. As manual checks are time-consuming and error prone, automating such tests is desirable.

In this paper, we describe further research directions on how to move onward with (automated) OOD testing. Our contribution is twofold:

- We demonstrate how to assess robustness of state-of-the-art AI/ML approaches in OOD situations using two data sets from natural language and image processing.
- We formulate open issues and possible directions for further research focusing on OOD detection, handling, and prevention in interactive AI systems with regard to methods, tools, and development processes.

We will first review related work (Sect. 2). Subsequently, we report findings from two case studies to evaluate the robustness of state-of-the-art AI/ML approaches with two openly available datasets from different application domains. We demonstrate how applications can be tested for OOD situations with the examples of natural language and image processing (Sect. 3). We then propose a research agenda for OOD tests in interactive AI systems and give first recommendations on how to solve these issues (Sect. 4). We conclude the paper with an outlook of future work (Sect. 5).

2 Related Work

In this section we will first review how AI/ML models can be tested in general, especially looking at methods and processes (Sect. 2.1). We then discuss approaches dealing with testing for OOD situations (Sect. 2.2).

2.1 AI/ML Testing

AI/ML techniques have plenty of use in software testing: Durelli et al. [8] report on a variety of use cases, among them generating test-cases, refining test-cases, and evaluating oracles. Such approaches focus on automating different aspects of software testing. However, besides regular software testing, using AI/ML in software components requires dedicated testing of the trained AI/ML models themselves, which is a whole different (and in some parts new) field of software testing. Such tests revolve around the underlying data foundation, the model's predictions, and the runtime behavior, especially in new environments.

Breck et al. [6] introduce a score for determining whether a given AI/ML component is ready for production. The score is built upon four categories of tests with several details: tests for features and data, tests for model development, tests for ML infrastructure, and monitoring tests for ML. While the first category seems related to our work at a first glance, the best fit is in the infrastructure category: here, Breck et al. [6] require automated evaluation of the model quality before the deployment. They specifically mention the degradation of model quality over several versions and we argue that data quality – and especially the test for OOD situations – are an important aspect to consider in this stage of the development cycle. Nevertheless, also the *tests for features and data* include aspects relevant to our work: For example, Breck et al. [6] require developers to use a schema for documenting their expectations of features and their value space (most notably for testing against unexpected or wrong values in the training data).

Riccio et al. [38] differentiate between different levels of testing for applications using AI/ML components: input testing (tests on the data used for training and inference), model testing (testing the AI/ML component in isolation), integration testing (testing the interplay of different system components), and system testing (testing the overall system). They also define *Data-Box Testing* as a new type of testing for AI/ML besides black-box and white-box testing. Such tests have access to training and test data and, in simple terms, use manipulations of the data to assess changes in the model's behavior. They also show that manipulation of input data (sometimes referred to as *mutation testing*) is a common approach to testing AI/ML components. Riccio et al. [38] compared and grouped some existing publications and their methods into categories. For example, they identified several approaches that already use *input mutation*. Input mutation means that new input possibilities are derived from modifying existing inputs. In particular, image manipulation, which we will also use in our case studies, is discussed here.

DeepMutation [32] also uses input mutation, more precisely image manipulation. The generated new inputs are given to trained models and the results are compared with those obtained on widely used data sets. Five different options are presented for source-level mutation: *Data Repetition, Label Error, Data Missing, Data Shuffle* and *Noise Pertubation*. *DeepMutation++* [25] is a successor to DeepMutation that allows not only to manipulate inputs, but also to manipulate the executed models. Via different starting points, such as a manipulation of the model weights, the functionalities of the software are also tested.

Tian et al. [47] recently released *LEAM*, which should help finding mutation faults based on real faults. For example, the correct syntax of the input is taken into account, so that the newly generated input is more realistic and closer to the original input. The tool outperforms existing technologies including DeepMutation in the areas of mutation testing, mutation-based test case prioritization, and mutation-based fault localization.

2.2 OOD Testing

OOD situations result from significant differences between the data used in training and inference [34]. In previous research, different methods have been proposed on how to increase the reliability of trained AI models to handle OOD situations (see, e.g., Salehi et al. [40]). Wang et al. [49] categorized these techniques into (1) data manipulation, (2) representation learning focusing on the conceptual abstraction, and (3) general learning strategies. *Data Augmentation* is a common step in deep learning pipelines that artificially manipulates training data and therefore adds more variance to the data set. The manipulation operations depend on the data type, e.g., image data can be cropped or color inverted [44] and in time series, jitter or time shifts can be added [26,27]. Such manipulations that also occur during real-world operation make the generated inputs become OOD situations.

But Data Augmentation is also constrained by aspects from the application domain. Training a model with too heavily manipulated data could lead to corrupted learnings and compromise the main objective of the system. Discussing constraints for Data Augmentation with stakeholders might become an early-phase activity in SE processes for AI-based applications (see, e.g., Amershi et al. [1] or Hesenius et al. [23]). Besides, previously unseen augmented data can also be used in the testing phase to validate the robustness of trained models before deployment.

Schelter et al. [42] introduce *JENGA*, a framework to automatically verify how errors in data impact a model's performance. Such frameworks can be used to conduct *What-If-Analyses* [11,12] that inspect AI/ML processing pipelines for their reaction when receiving faulty data.

A similar type of test with a slightly different goal are *Adversarial Examples* [46,52]. While the authors aim to misuse the model in a hostile manner and thus show weaknesses and potential attacks, our approach is more focused on ensuring proper UX. We also argue that our approach requires not only technical specialists, but also representatives from the application domain to define proper test scenarios.

3 Handling OOD Situations: Case Studies

OOD situations occur when an AI/ML model faces data that is somehow different from the data seen during training. In HCI, such situations can be quite common: users with a local dialect can use voice recognition or users with motor impairments can use applications based on movement, e.g., gestures. As a result, predictions may be unreliable. If the OOD situation is not detected before inference, the model will nevertheless return a result, maybe even with a certainty over the defined threshold. There is a chance that such results will be wrong and thus yield errors in subsequent processing, which in turn could lead to potentially unwanted behavior and unsatisfactory UX. Therefore, we have a clear need to test how robust models are towards OOD situations before deployment. Furthermore, developers should identify features especially prone to shifts in the data to mitigate problems systematically. However, the variations observable during runtime can result in a huge combinatorial space, thus it might be unfeasible to collect sufficient training and test data for all possible combinations. Furthermore, different problem types might lead to OOD situations: Some can result from the specifics of the application domain (e.g., the need to process words with special meaning) and others from pure technical aspects (e.g., delays in data synchronization can affect model performance [28]).

A common approach is *Data Augmentation*, i.e., adding additional information, sometimes synthetic, to the data. A large number of frameworks exist that support developers in adding synthetic manipulations to data. For example, Augraphy[1] can be used to augment images of documents with typical problems arriving from e.g., bad scan quality, and AugLy[2] supports over 100 modifications for image, video, text, and audio data. ImageNet is a common dataset for computer vision tasks and several augmented variants have been proposed (e.g., ImageNet-R [15], ImageNet-ReaL [4], ImageNet-A [18], or ObjectNet [3]). For augmenting Natural Language Processing (NLP) data, Li et al. [31] describe various techniques. Also, Zhou et al. [51] review various OOD datasets.

In the following sections, we describe two case studies on how possible OOD situations can be detected in common AI/ML application scenarios: text-based

---

[1] https://github.com/sparkfish/augraphy.
[2] https://github.com/facebookresearch/AugLy.

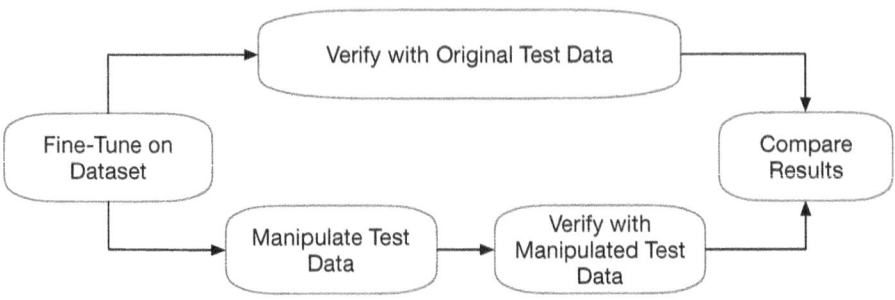

Fig. 1. General workflow for both case studies.

NLP (Sect. 3.1) and image processing (Sect. 3.2). For both case studies, we report the impact of manipulated data on the model prediction performance and compare it to the original test set performance. Figure 1 shows an overview of the general approach applied to both case studies. We argue that the general approach can be applied to any type of data, e.g., sensor data, time series data, or tabular data. Moreover, we focus on a single data source, but AI/ML projects will often combine data from multiple sources. Developers must thus explicitly take synchronization of data sources into account as it potentially introduces ODD situations at runtime if data synchronization differs from the training data [28].

3.1 Text-Based Natural Language Processing

Since the arrival of Transformers [48], NLP has received a lot of attention[3]. While LLMs are made available as large pre-trained models and are thus interesting for developers as a foundation for applications, they have also been been reported to be prone to attacks with *adversarial prompts* where attackers add suffixes to valid prompts and force the models to generate undesired results [52]. Besides such apparent misuse, LLMs need to deal with the variance that natural language entails: text in newspapers is different from text in social media. Internet culture has also given rise to various trends that build on natural language, e.g., using emojis instead of words.

Text-based NLP offers interesting capabilities for HCI, for example chatbots or sentiment recognition. Classic chatbots have to be specifically designed for their use case, i.e., expected user input (often called the *utterance*) and the connected aim (the *intent*) have to be predefined for each step in the dialog, which is why dedicated design approaches have been developed (an example can be found in Börsting and Hesenius [5]). That chatbots based on LLMs surpass classic chatbots in many aspects is currently showcased by applications such as ChatGPT[4], but they still require some sort of fine-tuning to solve specific tasks. Sentiment recognition is the task to derive user emotion from their written text,

---

[3] No pun intended.
[4] https://chat.openai.com.

Table 1. Manipulations for Leetspeak

A → 4	B → 8	E → 3	G → 6
L → 1	O → 0	S → 5	T → 7

which can be used in may ways in HCI, e.g., to enhance a chatbot's reaction or to analyze application reviews.

We will focus on sentiment recognition in social media for this case study. The case study shows how simple manipulations of texts from social media can distort an LLMs understanding and thus impact the application's results.

Text Data Set. The dataset *Coronavirus Tweets NLP - Text Classification*[5] contains about 4,000 tweets with messages about the coronavirus and the resulting crisis labeled into the following classes: *extremely negative, negative, neutral, positive*, and *extremely positive*. In this case study, we only use the text of the original tweet and the given label, all other features are ignored.

Data Preparation. We aimed for text manipulations as close as possible to what humans would write in social media, i.e., introducing additional symbols or making slight mistakes etc. We adapted the text processing in such a way that manipulated texts could still be understood by humans, thus avoiding the generation of gibberish that would only not resemble meaningful text anymore. We chose four manipulations for this case study:

- *No punctuation*: For space limited or intentionally short messages, people tend to leave out punctuation marks. Missing (or wrong) punctuation has been reported to decrease model performance [24], although more recent models have been reported to be more robust than older approaches [9].
- *Spelling errors*: This manipulation contains typing errors that are often corrected by auto-correction systems like scrambled letters.
- *Leetspeak (LEET/L33T/1337)*: Leetspeak is a slang originating in gaming communities. It replaces some letters with similar-looking digits. Leetspeak has been reported to decrease NLP performance [24].
- *Emojis*: Emojis are small images of emotions embedded in text.

In order to analyze the text manipulation, a simple setting was chosen, which performed well on the available data. Using an LLM, the tweets were classified according to their sentiment. Since five different classes are available, an existing BERT-based model (DistilBERT [41]) was fine-tuned on this data set. We then applied the aforementioned manipulations on the test set. For the Leetspeak manipulation, letters are replaced by similar looking digits as denoted in Table 1.

Table 2 contains the mapping between words and emojis that we used to manipulate the text data. In addition to widely used emojis, we explicitly added

---

[5] https://www.kaggle.com/datasets/datatattle/covid-19-nlp-text-classification.

Table 2. Pairs of replaced words and the emojis used

five emojis relevant to the dataset context. As we use tweets about the coronavirus pandemic, we added `corona/mask`, `vaccination`, `hospital`, `doctor`, and `virus`. We manipulated the text by introducing the corresponding emoji-code from the Unicode standard[6], as the tokenizer used must be able to tokenize the characters, which is not possible for the pictorial emojis.

Results. The effects of the various text manipulations were different. For the sentiment classification the ROC-AUC scores were calculated in order to compare the effects with the original data. With the raw data set, the fine-tuned DistilBERT model achieved a ROC-AUC score of 0.92. Removing punctuation marks (first manipulation) had an effect only in the fourth decimal place. Swapping the first two letters resulted in a major effect: the ROC-AUC score dropped to 0.57. In the Leetspeak manipulation, where certain letters were replaced by numerical digits, our fine-tuned model achieved a ROC-AUC score of 0.63, just slightly better than before. Replacing certain words with emojis also had a noticeable effect on the model's performance, but not to the same extent: the ROC-AUC score was 0.89.

The results show how simple manipulations caused our LLM performance to drop. It is conceivable that a chatbot might encounter such changes in words while interacting with, e.g., a customer, thus developers need to check their chatbots capabilities accordingly.

3.2 Images

Image processing is a common and widely adopted technique in various domains, e.g., in dentistry to detect tooth diseases [10,30] or in autonomous driving to detect obstacles [33]. In HCI, image processing techniques can be used for gesture recognition, activity detection, or also emotion recognition, just to name a few examples. But image processing is also prone to *Adversarial Examples*, i.e., even slight modifications invisible to the human eye impact model perfor-

---

[6] https://unicode.org/emoji/charts/full-emoji-list.html.

mance [46], which, conversely, suggests even worse results for potential defects that can happen at runtime.

For this case study, we will focus on image manipulations resembling natural occurences. For example, the sensors lens might have been touched and a fingerprint left as residue or weather conditions are bad, i.e., it is raining or foggy. Figure 2 shows some examples.

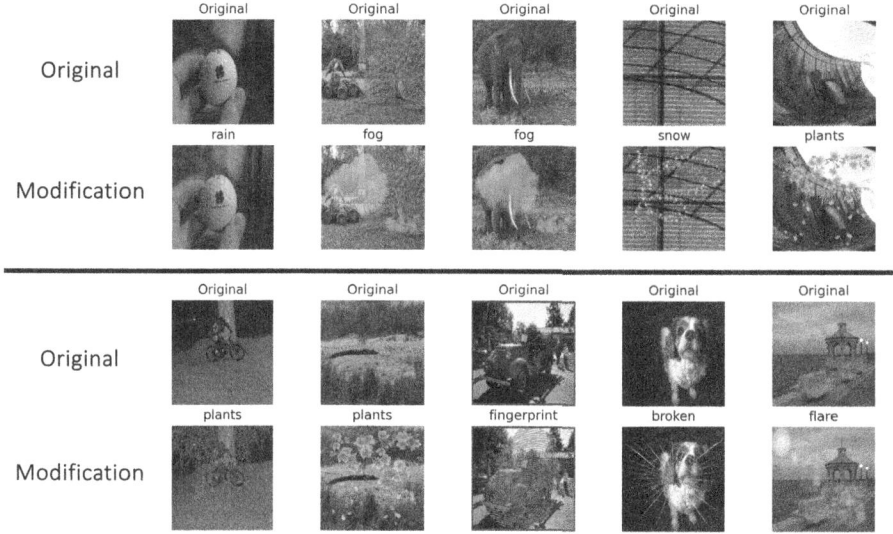

Fig. 2. Image Modifications

Image Data Set. We used the ImageNet-1k data set consisting of labeled images sorted into 1,000 object classes ranging from goldfishs to sewing machines. The data set is available[7] with pre-configured, non-overlapping splits consisting of 1.2 million training images, 50,000 validation images, and 100,000 test images. In our study, we applied several manipulations to 1,000 images randomly sampled from the ImageNet-1k validation data set. We used the validation set as the ImageNet-1k test set has no labels available (an approach inline with previous work such as Hendrycks et al. [16]).

Data Preparation. Extending the training set with manipulated images before training is a common step in deep learning pipelines. For example, Felsch et al. [10] improved the robustness of their caries detection AI model by including augmented images of teeth into their training data. The basic idea is forcing the

---

[7] https://huggingface.co/datasets/imagenet-1k.

model to learn the concept of interest (e.g., a tooth) instead of relying on additional information in the image (e.g., things that are surrounded by red colored skin are teeth). Common operations are flipping the image, changing the image's color or brightness, cropping away parts, or superimposing the image on other unrelated images showing everyday objects. Nevertheless, we argue that beside these basic manipulations, further more complex and realistic manipulations are needed depending on the use case to ensure the models' robustness during operation. Therefore, the selected manipulations should represent quality issues that may also be faced during real-world model operation. Our manipulations only augment a small part of the image, e.g., by adding a lens flare from the sun in the right part of the image, instead of adjusting the brightness of the whole image. We have chosen the following image manipulations:

- *Weather conditions*: Introduces artifacts from snow, fog, or rain into the image.
- *Lense Flare*: Introduces reflections into the image, resulting in overly bright patches.
- *Fingerprint*: Overlays parts of the image with a fingerprint like someone has touched the lens.
- *Broken Glass*: When a smartphone is dropped, the camera lens can be damaged with hairline cracks that results in artifacts in the image.
- *Falling Leaves*: The original image motif might be partly obscured by falling leaves.
- *Tree Branches*: Due to an unfavorable shooting angle, branches may get in the way of the camera.

In addition to text data, we evaluated the robustness of a recent computer vision model on an image classification task. We used a Vision Transformer (ViT) model openly available[8] that was pre-trained on the ImageNet-21k data set and afterwards fine-tuned on the ImageNet-1k data set consisting of 1,000 object classes. We applied the manipulations described before to 1,000 images randomly sampled from the ImageNet-1k validation data set. On a technical level, our manipulations are transparent image overlays that are added to the images in the validation data set. We augmented all images in the validation data set sample with our manipulations and used the vision transformer model to classify the main image motive.

After fine-tuning, we evaluated the model's performance with the following steps. First, we used the ViT model to classify the original images from the ImageNet-1k validation set sample without any augmentations and saved the five most likely class labels. We used the TOP-1 and TOP-5 accuracy as evaluation metrics and therefore focused on the label positions in the class label ranking and not on the probability of each label. Afterwards, we also classified the augmented image variants. Using the class labels from the validation set as ground truth, we calculate the TOP-1 and TOP-5 accuracy of the unmodified image and our 11 augmented image variants. When comparing the baseline

---

[8] https://huggingface.co/google/vit-base-patch16-224.

accuracy to the performance on augmented images, we can measure the impact of each augmentation on the predictive performance.

Results. We used the TOP-1 and TOP-5 accuracy metrics, which are commonly used for model evaluation on the ImageNet-1k dataset [39]. The TOP-1 accuracy measures how often the predicted class label with the highest probability equals the ground truth label. The TOP-5 accuracy works similar but compares how often the ground truth label is in the five most probable class label predictions.

The performance scores measured as TOP-1 accuracy are shown in Fig. 3a and the TOP-5 accuracy in Fig. 3b.

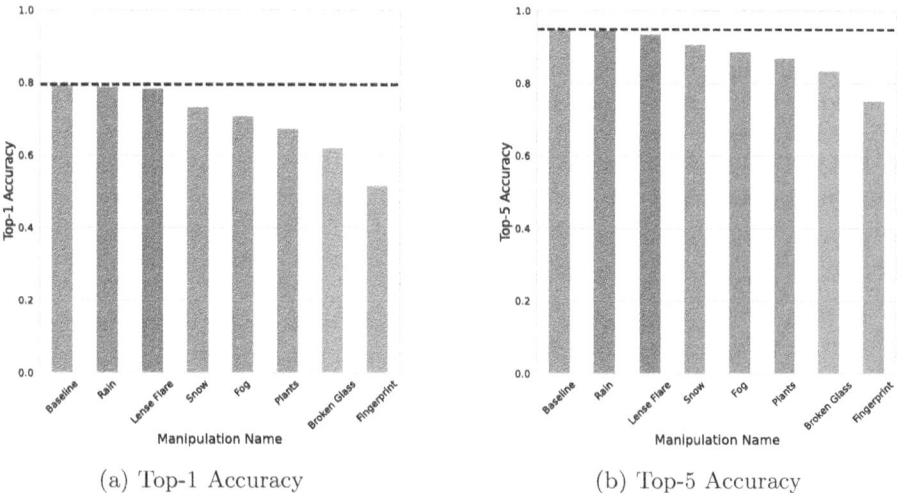

(a) Top-1 Accuracy (b) Top-5 Accuracy

Fig. 3. Performance metrics for the computer vision task

On the original unmodified images, the ViT model achieved a TOP-1 accuracy of 79.5%. When augmenting the images with rain and lens flare overlays, we observed only minor performance degradation (0.5%–1%). With added snow, the TOP-1 accuracy decreased noticeably by 6.1%. Adding fog and plant overlays as well as a broken glass overlay to the images resulted in performance scores from 8.5% to 17.3% worse than the baseline TOP-1 accuracy. Finally, the fingerprint overlay had the worst TOP-1 accuracy with 51.8% representing a difference by 27.7% compared to the unmodified baseline.

For the TOP-5 accuracy, the overall performance increases but the impact ranking of our modifications remains the same. On the unmodified image, the ViT achieved a TOP-5 accuracy of 94.8% and we observed almost the same score for images with the rain overlays. Lens flare artifacts decreased the TOP-5 accuracy by 1.3%, snow artifacts by 4%, and fog overlays by 6%. When augmenting the original image with leaves, we observed a TOP-5 accuracy of 87.3%

representing a decrease by 7.7%. The broken glass decreased the performance by 11.2% and finally, the fingerprint overlay classified our sample with a TOP-5 accuracy of 75.3% which is a decrease by 19.5%.

Again, we could observe severe drops in model performance for different manipulations. As image processing is quite common in HCI, e.g., to monitor users or as a means for gesture recognition, developers need to check their models accordingly. Some of these problems are also quite hard to identify for users – for example, hairline cracks are barely visible on lenses – making correct handling by the application more important.

4 Open Issues and Research Questions for Engineering Interactive Systems Embedding AI Technologies

OOD situations need to be explicitly considered when engineering systems embedding AI/ML components to prevent critical UX issues. In our case studies, we have demonstrated how to test the robustness of AI models to identify possible OOD situations and assess the impact on the model's performance. The case studies have confirmed that OOD situations are still relevant to modern AI models. Therefore, we describe open issues in OOD testing related to methods, tools and the development process of AI components in general and provide first hints on how to move onward.

Preventing OOD situations requires special attention across the whole AI development process: During Requirements Engineering, it might be discussed that events like broken lenses or network delays could lead to OOD situations. In the data collection phase, we could introduce data augmentation operations to detect and report OOD situations or to teach the model tolerances up to certain thresholds. In the model testing phase, we could use augmented data to check e.g., up to which threshold our model is robust to network delays and which effect greater delays have on our prediction performance.

In the following, we will outline open research questions for two key aspects of OOD testing in interactive applications: *detecting* OOD situations (Sect. 4.1) and *handling and preventing* OOD situations (Sect. 4.2). We will focus on the necessary methods, tools, and necessary adaptations to development processes.

4.1 Detecting ODD Situations

After model deployment, we need to detect when ODD situations occur. Trained models will try to reason about conceptually shifted data based on the previously learned world knowledge and output a valid estimation that might be incorrect from the view of a domain expert. In previous research, state-of-the-art transformer models were trained to detect concept shifts and report potential ODD risks [29]. Another detection method is to provide developers with visual clues about the data used during inference and to continuously compare the data distribution with the training data [45]. A further line of research formulates the detection of ODD situations as a problem of anomaly and outlier detection [17].

We envision an AI testing framework that enables domain experts to describe possible OOD scenarios in natural language and the framework will automatically generate test data that represents such scenarios, e.g., by applying manipulations as described in Sect. 3 or generating new data based on templates, e.g., synthetic gestures as proposed by Hesenius et al. [19]. Especially the latest advances in generative AI could facilitate the generation of realistic test data for tasks e.g., in computer vision or NLP. Furthermore, in Behavior-Driven Development (BDD), it is quite common to define tests in pseudo-natural language [50], which has also been applied to HCI [13,14,20,21].

Our experimental case study on NLP showcased the impact of modifications like emojis and Leetspeak on sentiment analysis. Abstracting from this special case, there is a need for methods, tools and dedicated considerations in the development process related to detecting OOD situations. Only if we detect that our model does not react as expected on input such as Leetspeak, we can apply appropriate countermeasures.

Methods. Detecting OOD situations requires integrating a new kind software testing into development processes. To get a better understanding how existing testing methods must be adapted, we envision more in-depth investigations on possibilities and limitations of current practices. To overcome the current limitations, we envision further methods to support OOD testers, leading to the following **research questions**:

- To what extent does the detection of OOD situations depend on the data type and application context?
- Can OOD recognition methods for the different data types be standardized?
- Can these approaches be applied in multi-modal settings (e.g., image data combined with voice data)?
- How does multi-modality affect an application's proneness to OOD situations?

To answer the raised questions, we recommend the following **actions**:

- Categorization of OOD detection methods for different data types
- Experiments to assess the performance of OOD detection methods on datasets with different and mixed data types
- Definition of possibilities for developers to rate the expected proneness of their application to OOD situations
- Definition of methods to identify features or combination of features most prone to OOD situations
- How can variations in user input, a typical cause for OOD, be identified during runtime?
- How can variations in user input be identified during development?

Tools. On a tooling level, we have the difficulty that OOD tests require input from domain experts, data scientists, and software engineers and so the tools must address needs of all involved stakeholders. As a consequence, we raise the following **research questions**:

- Which context data should be supplied for AI prompts?
- How and in which form should the context be supplied?
- Are there metrics that can provide initial indications of OOD situations?
- Can proneness to OOD derived from other information, e.g., requirements definitions?
- How can OOD situations be identified and discussed between team members?
- How can variations in user input be defined?

We envision the following **actions** for a more comprehensive investigation:

- Further experiments on the robustness of state-of-the-art AI models on datasets with different data types and mixed-up data types (e.g., images, texts, tabular data, time series data). To trigger OOD situations, it might work out to combine several data sets and, e.g., test general-purpose LLM models on highly specific medical or physics questions.
- Experiment on AI model prompt engineering for different data types, levels of detail and input formats of context information.
- Identify inference data metrics that might be helpful as initial OOD indicators.
- Develop notations for describing OOD situations, especially with regard to user variations.

Development Process. Processes define not only the steps to archive a certain goal, but also the responsibilities and necessary skills of the stakeholders involved. As described before, OOD testing requires input from different stakeholders but the responsibilities remain unclear. The following **research questions** might help to guide further research:

- How should responsibilities for OOD situations be distributed between stakeholders?
- Do development teams require additional skills to handle OOD situations?
- In which phase must activities for OOD testing be planned and executed?
- How can requirements for detecting OOD situations be defined?
- How can HCI-related OOD situations be highlighted and identified during development?

4.2 Handling and Preventing OOD Situations

If we know that our model ran into an OOD situation during inference, we need appropriate countermeasures. These situations can be sporadic, like the broken camera lens, or they can be constant over time, like a person with a rare regional dialect trying to use a voice assistant. With a report of the OOD situation and the usage context at hand, there are several options for resolving the situation. We might collect additional data and fine tune our model to handle the specific OOD situation like training our computer vision model on images with lense cracks inside the image to be more robust against failures. The cracks might be added artificially as part of data augmentation or new data could be collected using different kinds of cracked lenses. Beside fine tuning a generalized model, we could also train a dedicated model for our specific use case, then let both models predict the input data and decide for the output with the higher probability or based on a voting mechanism. On the upside, this might lead to highly accurate predictions as we have a good performance on normal as well as corrupted data. On the downside, the more issues we might face in practice the more models we need to train from ground up to handle all such situations resulting in high computational training demand. As discussed earlier, it may also be possible to identify potential OOD risks in Requirements Engineering, and therefore we could collect or artificially generate data representing such situations. Testing the robustness levels of trained models in these situations may also be a viable option, and has been tested, for example, by Ovadia et al. [37] on image data.

In our NLP case study, the problems arriving from OOD situations could be handled by tools to automatically replace all emojis with a word representation or a use-case specific text pre-processing and transformation step in the AI/ML model development process. More generally, we suggest the following **research questions and actions** for handling and prevention of ODD situations with regard to methods, tools, and the development process.

Methods. Methodical guidance for people involved in AI systems engineering is required to leverage best practices in decisions related to OOD testing. While detecting OOD situations is reactive, dealing with and preventing OOD situations is proactive. We encourage further research focusing on the **research questions** below:

– How to decide what new data to collect to address the OOD situation?
– Does the integration of new data have an impact on the prediction quality of normal data?
– How can we determine an end-of-life point for models when the concept shift has become too large?
– How to decide if a model should be fine-tuned for a use case, or if we should rather develop our own specific model for it?
– Why did the model return an incorrect result and which data should we collect to correct false learnings?
– How can applications adapt to users to prevent OOD situations (i.e., integrating individual data to learn about users nuances)?

To answer our questions raised, we propose the following **actions**:

– Experiments on how well AI models can be used to generate more training samples (e.g., modification from a dialect recording or synthetic gestures)

- Experiments on how well AI models trained for different data contexts can be combined to overcome OOD situations in at the intersection of contexts (e.g., combine German and English speech recognition to understand speakers mixing up both languages)
- Formulation of Requirements Engineering workshop or interview guidelines on how to elicit potential OOD risks before model training

Tools. OOD testing can also be supported by tools and frameworks usable by developers as well as domain experts to guide OOD prevention. This also requires a deeper understanding of the targeted user group to develop adequate HCI methods and create the desired UX, which in turn must be tested. We raise the following **research questions**:

- How can data collection be supported if there are high costs or effort associated with collection (e.g., physiologic medical data, acquisition of speakers of a particular regional dialect)?
- How can we debug wrong predictions for non-image data and trace the decisive part of the input data?
- How can we generate synthetic data that includes specific variations of different user groups?

To answer our questions raised, we propose the following **actions**:

- Development of a framework that allows generating training data based on natural language scenario specifications
- Development of data type specific tools to trace the AI model output to the most decisive input data

Development Process. From a broader perspective, the process of engineering AI embedding software requires dedicated roles, skills and responsibilities [23]. For OOD prevention, we formulate the following **research questions**:

- Which skills are required to prevent OOD situations already before the model deployment?
- Which stakeholder(s) should be responsible for OOD handling and prevention?

Our questions could be addressed with the following **actions**:

- Integration of OOD considerations into each step of AI engineering process models
- Observation of developers accompanied by interviews

5 Conclusion and Future Work

Ensuring reliable AI models will be the key to a satisfying UX. This is particularly true if the AI is connected to processing data relevant for HCI. With two case studies in the ares of NLP and image processing, we demonstrated how the reliability of AI components decreases if the model is confronted with distorted input data. Such OOD situations can threaten the UX but are hard to control for developers. We argue that new test automation systems are necessary to support software developers when testing applications for potential defects in OOD situations. We defined open research questions regarding the detection, handling, and prevention of OOD situations that threaten the reliability of AI models in production. To answer these questions, we gave first directions on the next steps. Of special interest are benchmarks on the robustness of state-of-the-art models in OOD situations concerning various data types as well as guidelines and tools on how to integrate OOD considerations into the overall AI systems engineering process.

Beside our research agenda on OOD testing, we envision further efforts in the general field of AI testing and debugging. Testing not only the robustness of AI models itself, but also the robustness and impact of changes in the data domain might be of interest before productive model deployment. One direction could be techniques to identify maximum acceptable data mutation thresholds beyond which model performance decreases significantly. Furthermore, developers should be able to identify features that are more prone to OOD situations than others. In addition, individual user differences can also lead to OOD situations. Therefore, dedicated usage profile-oriented testing approaches may be another direction to ensure a high quality UX (see, e.g., Hesenius et al. [19]).

References

1. Amershi, S., et al.: Software engineering for machine learning: a case study. In: 2019 IEEE/ACM 41st International Conference on Software Engineering: Software Engineering in Practice (ICSE-SEIP), pp. 291–300. IEEE (2019). https://doi.org/10.1109/ICSE-SEIP.2019.00042
2. Anthony, L., Brown, Q., Nias, J., Tate, B., Mohan, S.: Interaction and recognition challenges in interpreting children's touch and gesture input on mobile devices. In: ITS 2012, pp. 225–234. Association for Computing Machinery, New York (2012). https://doi.org/10.1145/2396636.2396671
3. Barbu, A., et al.: ObjectNet: a large-scale bias-controlled dataset for pushing the limits of object recognition models. In: Advances in Neural Information Processing Systems, vol. 32 (2019)
4. Beyer, L., Hénaff, O.J., Kolesnikov, A., Zhai, X., Oord, A.V.D.: Are we done with ImageNet? (2020). https://doi.org/10.48550/arXiv.2006.07159. arXiv preprint
5. Börsting, I., Hesenius, M.: Towards a systematic approach for chatbot development in digital work environments. In: Klumpp, M., Ruiner, C. (eds.) Digital Supply Chains and the Human Factor. LNL, pp. 79–94. Springer, Cham (2021). https://doi.org/10.1007/978-3-030-58430-6_5

6. Breck, E., Cai, S., Nielsen, E., Salib, M., Sculley, D.: The ML test score: a rubric for ML production readiness and technical debt reduction. In: 2017 IEEE International Conference on Big Data (BIG DATA), pp. 1123–1132 (2017). https://doi.org/10.1109/BigData.2017.8258038
7. Brie, P., Burny, N., Sluÿters, A., Vanderdonckt, J.: Evaluating a large language model on searching for gui layouts. Proc. ACM Hum.-Comput. Interact. **7**(EICS) (2023). https://doi.org/10.1145/3593230
8. Durelli, V.H.S., Durelli, R.S., Borges, S.S., Endo, A.T., Eler, M.M., Dias, D.R.C., Guimarães, M.P.: Machine learning applied to software testing: a systematic mapping study. IEEE Trans. Reliab. **68**(3), 1189–1212 (2019). https://doi.org/10.1109/TR.2019.2892517
9. Ek, A., Bernardy, J.P., Chatzikyriakidis, S.: How does punctuation affect neural models in natural language inference. In: Proceedings of the Probability and Meaning Conference (PaM 2020), pp. 109–116. Association for Computational Linguistics, Gothenburg (2020). https://aclanthology.org/2020.pam-1.15
10. Felsch, M., et al.: Detection and localization of caries and hypomineralization on dental photographs with a vision transformer model. NPJ Digit. Med. **6**(1), 198 (2023). https://doi.org/10.1038/s41746-023-00944-2
11. Grafberger, S., Groth, P., Schelter, S.: Towards data-centric what-if analysis for native machine learning pipelines. In: Proceedings of the 6th Workshop on Data Management for End-to-End Machine Learning, DEEM 2022. Association for Computing Machinery, New York (2022). https://doi.org/10.1145/3533028.3533303
12. Grafberger, S., Groth, P., Schelter, S.: Automating and optimizing data-centric what-if analyses on native machine learning pipelines. Proc. ACM Manag. Data **1**(2) (2023). https://doi.org/10.1145/3589273
13. Griebe, T., Hesenius, M., Gesthüsen, M., Gruhn, V.: Test automation for speech-based applications. In: New Trends in Software Methodologies, Tools and Techniques: Proceedings of the Fifteenth SoMeT_16, pp. 85–100. IOS Press (2016). https://doi.org/10.3233/978-1-61499-674-3-85
14. Griebe, T., Hesenius, M., Gruhn, V.: Towards automated UI-tests for sensor-based mobile applications. In: Fujita, H., Guizzi, G. (eds.) SoMeT 2015. CCIS, vol. 532, pp. 3–17. Springer, Cham (2015). https://doi.org/10.1007/978-3-319-22689-7_1
15. Hendrycks, D., et al.: The many faces of robustness: a critical analysis of out-of-distribution generalization. In: Proceedings of the IEEE/CVF International Conference on Computer Vision, pp. 8340–8349 (2021)
16. Hendrycks, D., Liu, X., Wallace, E., Dziedzic, A., Krishnan, R., Song, D.: Pretrained transformers improve out-of-distribution robustness. In: Proceedings of the 58th Annual Meeting of the Association for Computational Linguistics, pp. 2744–2751. Association for Computational Linguistics (2020). https://doi.org/10.18653/v1/2020.acl-main.244
17. Hendrycks, D., Mazeika, M., Dietterich, T.: Deep anomaly detection with outlier exposure. In: International Conference on Learning Representations (2019)
18. Hendrycks, D., Zhao, K., Basart, S., Steinhardt, J., Song, D.: Natural adversarial examples. In: Proceedings of the IEEE/CVF Conference on Computer Vision and Pattern Recognition, pp. 15262–15271 (2021)
19. Hesenius, M., Book, M., Gruhn, V.: Test automation for gesture-based interfaces. In: HCI Engineering 2019 – Methods and Tools for Advanced Interactive Systems and Integration of Multiple Stakeholder Viewpoints. CEUR Workshop Proceedings (2019). http://ceur-ws.org/Vol-2503/paper1_5.pdf

20. Hesenius, M., Griebe, T., Gries, S., Gruhn, V.: Automating UI tests for mobile applications with formal gesture descriptions. In: Proceedings of the 16th International Conference on Human-Computer Interaction with Mobile Devices and Services, MobileHCI 2014, pp. 213–222. ACM, New York (2014). https://doi.org/10.1145/2628363.2628391
21. Hesenius, M., Griebe, T., Gruhn, V.: Towards a behavior-oriented specification and testing language for multimodal applications. In: Proceedings of the 2014 ACM SIGCHI Symposium on Engineering Interactive Computing Systems, EICS 2014, pp. 117–122. Association for Computing Machinery, New York (2014). https://doi.org/10.1145/2607023.2610278
22. Hesenius, M., Schwenzfeier, N., Meyer, O., Gruhn, V.: On the uncertainty in IoT-enabled business processes using artificial intelligence components. In: 2022 International Conference on Service Science (ICSS), pp. 80–87. IEEE, Zhuhai (2022). https://doi.org/10.1109/ICSS55994.2022.00021
23. Hesenius, M., Schwenzfeier, N., Meyer, O., Koop, W., Gruhn, V.: Towards a software engineering process for developing data-driven applications. In: 2019 IEEE/ACM 7th International Workshop on Realizing Artificial Intelligence Synergies in Software Engineering (RAISE), pp. 35–41. IEEE, Montreal (2019). https://doi.org/10.1109/RAISE.2019.00014
24. Hofer, N., Schöttle, P., Rietzler, A., Stabinger, S.: Adversarial examples against a BERT ABSA model – fooling BERT with L33T, Misspellign, and punctuation,. In: Proceedings of the 16th International Conference on Availability, Reliability and Security, ARES 2021. Association for Computing Machinery, New York (2021). https://doi.org/10.1145/3465481.3465770
25. Hu, Q., Ma, L., Xie, X., Yu, B., Liu, Y., Zhao, J.: DeepMutation++: a mutation testing framework for deep learning systems. In: 2019 34th IEEE/ACM International Conference on Automated Software Engineering (ASE), pp. 1158–1161 (2019). https://doi.org/10.1109/ASE.2019.00126
26. Iglesias, G., Talavera, E., González-Prieto, Á., Mozo, A., Gómez-Canaval, S.: Data augmentation techniques in time series domain: a survey and taxonomy. Neural Comput. Appl. **35**(14), 10123–10145 (2023). https://doi.org/10.1007/s00521-023-08459-3
27. Iwana, B.K., Uchida, S.: An empirical survey of data augmentation for time series classification with neural networks. Plos One **16**(7) (2021). https://doi.org/10.1371/journal.pone.0254841
28. Klumpp, M., et al.: Driving big data – integration and synchronization of data sources for artificial intelligence applications with the example of truck driver work stress and strain analysis. In: International Conference on Information Systems (ICIS) 2022 proceedings (2022). https://aisel.aisnet.org/icis2022/data_analytics/data_analytics/3
29. Koner, R., Sinhamahapatra, P., Roscher, K., Günnemann, S., Tresp, V.: OODformer: out-of-distribution detection transformer. In: 32nd British Machine Vision Conference 2021, BMVC 2021, Online, 22–25 November 2021, p. 223. BMVA Press (2021). https://www.bmvc2021-virtualconference.com/assets/papers/1391.pdf
30. Kühnisch, J., Meyer, O., Hesenius, M., Hickel, R., Gruhn, V.: Caries detection on intraoral images using artificial intelligence. J. Dent. Res. **101**(2), 158–165 (2022). https://doi.org/10.1177/00220345211032524
31. Li, B., Hou, Y., Che, W.: Data augmentation approaches in natural language processing: a survey. AI Open **3**, 71–90 (2022). https://doi.org/10.1016/j.aiopen.2022.03.001

32. Ma, L., et al.: DeepMutation: mutation testing of deep learning systems. In: 2018 IEEE 29th International Symposium on Software Reliability Engineering (ISSRE), pp. 100–111 (2018). https://doi.org/10.1109/ISSRE.2018.00021
33. Mao, J., Shi, S., Wang, X., Li, H.: 3D object detection for autonomous driving: a comprehensive survey. Int. J. Comput. Vision **131**(8), 1909–1963 (2023). https://doi.org/10.1007/s11263-023-01790-1
34. Moreno-Torres, J.G., Raeder, T., Alaiz-Rodríguez, R., Chawla, N.V., Herrera, F.: A unifying view on dataset shift in classification. Pattern Recogn. **45**(1), 521–530 (2012). https://doi.org/10.1016/j.patcog.2011.06.019
35. Mozaffari, M.A., Zhang, X., Cheng, J., Guo, J.L.: GANSpiration: balancing targeted and serendipitous inspiration in user interface design with style-based generative adversarial network. In: Proceedings of the 2022 CHI Conference on Human Factors in Computing Systems, CHI 2022. Association for Computing Machinery, New York (2022). https://doi.org/10.1145/3491102.3517511
36. Ojeda-Castelo, J.J., Capobianco-Uriarte, M.D.L.M., Piedra-Fernandez, J.A., Ayala, R.: A survey on intelligent gesture recognition techniques. IEEE Access: Pract. Innov. Open Solut. **10**, 87135–87156 (2022). https://doi.org/10.1109/ACCESS.2022.3199358
37. Ovadia, Y., et al.: Can you trust your model's uncertainty? Evaluating predictive uncertainty under dataset shift. In: Wallach, H., Larochelle, H., Beygelzimer, A., d' Alché-Buc, F., Fox, E., Garnett, R. (eds.) Advances in Neural Information Processing Systems, vol. 32. Curran Associates, Inc. (2019). https://proceedings.neurips.cc/paper_files/paper/2019/file/8558cb408c1d76621371888657d2eb1d-Paper.pdf
38. Riccio, V., Jahangirova, G., Stocco, A., Humbatova, N., Weiss, M., Tonella, P.: Testing machine learning based systems: a systematic mapping. Empir. Softw. Eng. **25**(6), 5193–5254 (2020). https://doi.org/10.1007/s10664-020-09881-0
39. Russakovsky, O., et al.: ImageNet large scale visual recognition challenge. Int. J. Comput. Vision **115**(3) (2015). https://doi.org/10.1007/s11263-015-0816-y
40. Salehi, M., Mirzaei, H., Hendrycks, D., Li, Y., Rohban, M.H., Sabokrou, M.: A unified survey on anomaly, novelty, open-set, and out of-distribution detection: solutions and future challenges. Trans. Mach. Learn. Res. (2022)
41. Sanh, V., Debut, L., Chaumond, J., Wolf, T.: DistilBERT, a distilled version of BERT: smaller, faster, cheaper and lighter (2020). https://doi.org/10.48550/arXiv.1910.01108. arXiv preprint
42. Schelter, S., Rukat, T., Biessmann, F.: JENGA: a framework to study the impact of data errors on the predictions of machine learning models. In: EDBT 2021 Industrial and Application Track (2021)
43. Shaw, A., Anthony, L.: Analyzing the articulation features of children's touchscreen gestures. In: Proceedings of the 18th ACM International Conference on Multimodal Interaction, ICMI 2016, pp. 333–340. Association for Computing Machinery, New York (2016). https://doi.org/10.1145/2993148.2993179
44. Shorten, C., Khoshgoftaar, T.M.: A survey on image data augmentation for deep learning. J. Big Data **6**(1), 1–48 (2019). https://doi.org/10.1186/s40537-019-0197-0
45. Song, D., Wang, Z., Huang, Y., Ma, L., Zhang, T.: DeepLens: interactive out-of-distribution data detection in NLP models. In: Proceedings of the 2023 CHI Conference on Human Factors in Computing Systems, pp. 1–17 (2023). https://doi.org/10.1145/3544548.3580741
46. Szegedy, C., et al.: Intriguing properties of neural networks. In: 2nd International Conference on Learning Representations, ICLR 2014 (2014)

47. Tian, Z., Chen, J., Zhu, Q., Yang, J., Zhang, L.: Learning to construct better mutation faults. In: Proceedings of the 37th IEEE/ACM International Conference on Automated Software Engineering, ASE 2022, Association for Computing Machinery, New York (2023). https://doi.org/10.1145/3551349.3556949
48. Vaswani, A., et al.: Attention is all you need. In: Guyon, I., et al. (eds.) Advances in Neural Information Processing Systems, vol. 30. Curran Associates, Inc. (2017)
49. Wang, J., et al.: Generalizing to unseen domains: a survey on domain generalization. IEEE Trans. Knowl. Data Eng. **35**(8), 8052–8072 (2023). https://doi.org/10.1109/TKDE.2022.3178128
50. Wynne, M., Hellesoy, A., Tooke, S.: The Cucumber Book: Behaviour-Driven Development for Testers and Developers. Pragmatic Bookshelf (2017)
51. Zhou, K., Liu, Z., Qiao, Y., Xiang, T., Loy, C.C.: Domain generalization: a survey. IEEE Trans. Pattern Anal. Mach. Intell. **45**(4), 4396–4415 (2023). https://doi.org/10.1109/TPAMI.2022.3195549
52. Zou, A., Wang, Z., Kolter, J.Z., Fredrikson, M.: Universal and transferable adversarial attacks on aligned language models (2023). https://doi.org/10.48550/arXiv.2307.15043. arXiv preprint

Doctoral Consortium EICS 2023

End User Development for Extended Reality

Valentino Artizzu(✉)

Department of Mathematics and Computer Science, University of Cagliari,
Cagliari, Italy
`valentino.artizzu@unica.it`

Abstract. This paper presents the early work of my PhD, which revolves around finding solutions to support end users in the process of the creation of Extended Reality experiences, without them having programming experience or 3D modelling skills. This approach overcomes the limitation of the end user users not able to develop their projects without the help of an expert developer. Current work is defining a system in the context of Vocational Training for creating procedures that work as a training or guidance environment, depending on the Extended Reality device, and an execution engine capable of adapting the procedure depending on the context of the environment of the user. Current work is also being showcased, where the interface of the system has been developed following the insights from the dedicated workshop and the results are tested with expert users.

Keywords: Extended Reality · Mixed Reality · Augmented Reality · Virtual Reality · End User Development · Natural Language · Rule System · Event-Condition-Action Rules · Vocational Training

1 Introduction

In recent years, consumer Extended Reality (XR) devices (Virtual, Augmented and Mixed Reality - VR, AR, MR in short) have become more accessible and popular. They offer immersive and fun experiences, mainly for gaming purposes, but they also proved to be effective in other fields, ranging from educational and learning experiences to tools that assist the user to finish work assignments. As with other technologies in the past, such as the Internet of Things (IoT) and web content management, the more widely adopted a technology becomes, the more users demand better control over creating content.

As of now, end users cannot handle the development of entire XR experiences, since they require a team of experts in 3D modelling, code development, design etc. [18,19]. Moreover, current development cycles are not suitable for end users with evolving needs over time, since the involvement of a professional developer is not always feasible.

The goal of this PhD research is to find solutions to support end users in creating XR experiences, without requiring programming experience or 3D modeling skills. The resulting idea is to create environments that allow the creation of immersive configurations that show the user what they can do, using "what you see is what you get" techniques, that can enable the definition of behaviours for Extended Reality experiences, through the use of rules expressed in natural language, and with XR interfaces that can help the user in creating the rules without ever leaving the headset. Finally, the results of this process will be validated through field testing to verify the effectiveness of the idea. In particular, my approach envisions a thorough investigation of all potential solutions after researching current literature, identifying potential points of improvement, and developing novel approaches for filling the gaps found in the field.

My most recent research defines a system for creating procedures that function as training or guidance environments, depending on the extended reality (XR) device, and an execution engine that can adapt the procedure to the environment of the user. This system is motivated by the need to provide end users with direct access to the experience creation and development process in the vocational training domain, where trainees learn from procedures.

In the next sections the current work will be discussed, starting from the current state of the art.

2 Related Work

In this section relevant work linked to this PhD research will be defined, outlining the current limitations of the literature.

2.1 End User Development for Extended Reality

My PhD research falls within the scope of the End User Development (EUD) approaches. The definition Lieberman et al. provided in their work has been used for defining the idea: "EUD can be defined as a set of methods, techniques, and tools that allow users of software systems, who are acting as non-professional software developers, at some point to create, modify, or extend a software artifact" [13]. This concept brings the end user to a position where he can be empowered "to develop and adapt systems themselves" [13]. In this work EUD has been applied in the context of the Extended Reality. Extended Reality is an umbrella term that includes Virtual Reality ("technology that provides almost real and/or believable experiences in a synthetic or virtual way" [10]), Augmented Reality ("a system that enhances the real world by superimposing computer-generated information on top of it" [10]), and Mixed Reality ("a continuum that includes both Virtual Reality (VR) as well as Augmented Reality (AR)" [10]). According to the literature most of the currently available work for XR experiences is limited to the definition of static scenes or multimedia content overlays.

Fungus [9] is open-source software for developing visual narratives that operate as an extension for Unity3D. It employs flowcharts to construct visual novels, but they require a large amount of screen space and reduce the end user's comprehension [12].

Eroglu et al. [7], instead, presented a VR authoring system for testing and automation of autonomous vehicles. The authors implemented an adapted 2D interface for enabling the user to add, modify and remove pieces of roads in a 3D environment. This tool, despite implementing an interesting set of controls and elements for authoring scenarios for autonomous cars, is highly specialised on the designed task, and cannot be used in a generic context, and my work aims to explore that field.

VREUD [22] is a web-based authoring tool for creating VR environments using a 2D interface. The tool enables users for the creation of scenes that can be customised using a 3D representation of the environment, and creation of behaviours in a similar manner to the previous work of me and my research group (presented in Sect. 2.2). Despite being a very interesting approach on giving the user the ability to create their own environments, and having many common points with the goals of my PhD research (the creation of environments without the use of code), this project differs from it in an important aspect, as the aim of my research is to create experiences that can be defined and executed without any external tools provided alongside the project itself.

Fanni et al. [8] proposed PAC-PAC, an authoring tool for creating point-and-click games using a Web interface, 360° photos and videos as scenes, and the definition of the behaviour made through natural-language, Event-Condition-Action rules. Unfortunately, even if a VR version mode was in place, the system was limited to first-person point and click games, with user interaction solely focused on solving puzzles, and not exploring a generic XR environment.

2.2 Published Work on End User Development for Extended Reality

From an analysis of the previous works, it can be seen that there was the possibility to design a system that could allow an end user developer to design an XR experience without being constrained to a single scope, while being immersed on the environment itself. To achieve that, ECARules4All [1] was developed, sharing its roots with PAC-PAC: the same Rule system guidelines has been ported and improved into a VR-focused derivative. In this case, the project extended the foundations of PAC-PAC to accommodate more complex interactions in full VR environments, while also defining the roles participating in the environment development: the Template Builders (expert developers who can create the templates, almost-complete VR environments, open to end user configurations), the End User Developers (users with average skills in computer use and VR that configure the templates to get the final version of the VR environment) and the Final Users (the final consumers of the experience).

2.3 End User Development for Vocational Training in XR Environments

As my work focuses on End User Development for XR, it has been possible, while defining the latest project, to leverage the same knowledge gathered for the ECARules4All project, and combine it with an investigation in the vocational training field.

Vocational training is a term that defines "education programmes that are designed for learners to acquire the knowledge, skills and competencies specific to a particular occupation, trade, or class of occupations or trades" [17].

Looking at the literature, many contributions can be found, for example work related to welding training [16] and construction management [14], proving that learning in an immersive environment can be beneficial to the end user. Unfortunately, from an end user point of view, none of these contributions gave the possibility to customise the experience once it was built by the developer, defining a gap the proposed work wants to fill.

Regarding vocational training using AR, Chiang et al. [4] conducted a systematic review on this matter, and they found that AR training has positive effects on vocational training outcomes.

In literature there are a multitude of examples that show that vocational training can also be adopted to VR applications. Wolfartsberger and Niedermayr [20] and Cassola et al. [3] exploit the idea of procedure authoring by demonstration in order to create the tutorial to be given to the trainees. The former discussed the potential benefits of VR in industrial training applications, specifically in assembly procedures and maintenance tasks. Their work presented a concept to simplify the authoring process of content with additional focus on animating assembly procedures. This is achieved through a "authoring-by-doing" approach where trainees can watch recorded actions performed by an expert on the VR environment and interactively mimic the set of instructions, with the goal of accelerating the content creation process for industrial VR-learning scenarios. The latter created a similar system that creates "choreographies" for each step to do on a 3D model, to be later viewed by the trainee during the training execution. The trainees can ask for a replay of a single step, if needed, and after the training has been completed the trainer can view the performance of the apprentice using the same recording system used during the authoring mode. Unfortunately, despite the interesting idea of showing how the trainee must interact with the scenario in order to achieve his goal, these works focused only on assembly processes, and my work focuses more on more broad aspects of training and customisation.

Wu et al. [21] proposed a framework to define a consistent set of vocabulary used throughout the design, implementation, and analysis of Guided User Tasks (GUTasks) in a VR embodied experience. Results show that the GUsT-3D framework and corresponding workflow implementation integrate well into the Designer's process to create, manage, and analyze VR embodied experiences of target users. Unfortunately they only completed a formative study, and it would be interesting to apply a similar approach to the current project, and have a better vision of real-world use case results.

XOOM [11] is a tool designed for non-ICT people that lets them create web-based VR applications using 360° videos and superimposing content into the virtual scenes. XOOM shares the idea to enable end users in the creation of immersive content is shared with ECARules4All [1], and the authors of the paper envisioned their tool to be used, among the variety of possibilities, as vocational training software, which is the objective of my current work, with the expansion of the concept using complete VR experiences. This expansion will take place in a similar fashion to what ECARules4All did, but focusing more on the process of generalising the training definition instead of creating universal rules for scene interaction.

In this case, after having analysed the state of the art from the perspective of the XR and vocational training, it can be seen that very few works try to give the possibility to the end user to define the behaviour of XR experiences without resorting to the expertise of a professional team. Moreover, there is no proper way to do so with XR vocational training experiences in particular, since usually the process of creating them is monolithic, and updating them is not feasible by the end user. As a potential solution for these issues, my current work will be presented in the next section, as it aims to be a possible way forward towards the realisation of this goal.

3 Current State of Research

The proposed work envisions the creation of a system that lets the end user developer define vocational training procedures with the aid of a dedicated authoring tool, allowing end users in the development process itself, without the need of an expert developer for the whole creation process of the final result. This system, moreover, has the capacity to be migrated to other platforms (i.e. any device that supports the implementation of it across the XR spectrum - training VR experience, guidance tool in AR). The work also proposes on the creation of an execution engine, that reads the procedure steps and translates it into XR interactions in real-time. The uniqueness of this system is that it also listens to the environment of the user, and adapts the modality (e.g. touch, controller click, voice, gaze-and-commit, etc.) of the interaction depending on the current context. This feature is particularly useful for simulating various real-world issues that can arise during the in-world execution of the task, and by doing so it is possible to train the novice worker to perform different actions for executing the same step in a procedure. This work presents a novel system that automatically suggests a modality for execution depending on the user's context, and a user interface that guides the user through the steps of a procedure in a clear and understandable way. Details of the workshop that produced the framework for the interface and the later expert evaluations are in Sect. 4. In this project, we can define different roles that interact with the system in different ways:

- Expert developers are users with good programming skills and proficiency in game programming platforms who can build XR applications. These users will be able to implement the components of the system inside an environment and enable end user developers to create the procedure steps;
- End User Developers are users without programming skills who require the expertise of a developer to implement the system proposed in this work. However, they are able to create a procedure using the authoring tool provided by the package. Once the procedure has been created, it can be distributed so that end users can engage with them;
- End Users are the users who will interact with the environment made by the expert developer and configured by the end user developer for learning how to do a task.

3.1 Methodology

To build this kind of support, the definition of the project is as follows (as also shown in Fig. 1):

- An authoring tool capable of creating the procedure files ready to be used or ported to other devices. This tool will be an immersive user interface where the end user developer will define the procedure steps, modalities, the target for the step, and step description data by interacting in the scene itself;
- A Procedure Engine that, given a complete procedure file, reads its base steps and advances in its execution, using a user-interaction-based method (e.g. when the user clicks on a button, the task is done and changes to the next one);
- A Context Engine that, based on the context of the environment, read through dedicated sensors, keeps track of the possible issues currently detected (e.g. loud environment, unrecognisable user voice, user preferences, etc.) and swaps the current modality in real-time, during the current task execution, or on task change, if the original next step cannot be executed with the current context.

To ensure the compatibility of the procedure with various devices, the project will include the Procedure Engine and the Context Engine as a part of a readily importable package that developers can easily integrate into their projects. The package will also include the authoring tool, but its inclusion is optional as it is not essential for the core functionality to function effectively. The concept of procedure portability facilitates the creation of reusable procedures that can be seamlessly integrated into diverse environments without the need for replication or full-scale sharing. Additionally, the design of the package as a modular unit rather than a complete environment allows developers to implement it in various systems while enabling compatibility with different XR devices. This flexibility empowers end users to freely choose which device best suits their needs and preferences when interacting with the system. To expedite implementation, the package will provide an example interface that can be utilised directly for rapid deployment of the various modes of the system. Furthermore, a set of APIs will be included to adapt the data from the available sensors within the target device into a format suitable for the Context Engine. To implement the project, the authoring tool must enable end user developers to define the steps and alternatives for the procedure using the immersive authoring interface. The required data for executing a step in the environment includes the target object, text information on the step and modality, the modality to be used for the interaction, and context information that triggers a change in the environment (e.g., if the environment is noisy, the "voice" modality must be changed to "touch"). The package is also expected to incorporate a mechanism for providing feedforward [6] to end users. Specifically, the provided user interface for procedure execution will display the information defined at procedure creation in a way that presents indirect feedforward to the user (e.g. if the user performs the modality "touch to the object "Ball", the step will be completed). Moreover, users will have the option to override the decisions of the system, if they so choose, by interacting with the user interface itself. An example of this interface can be seen in Fig. 2.

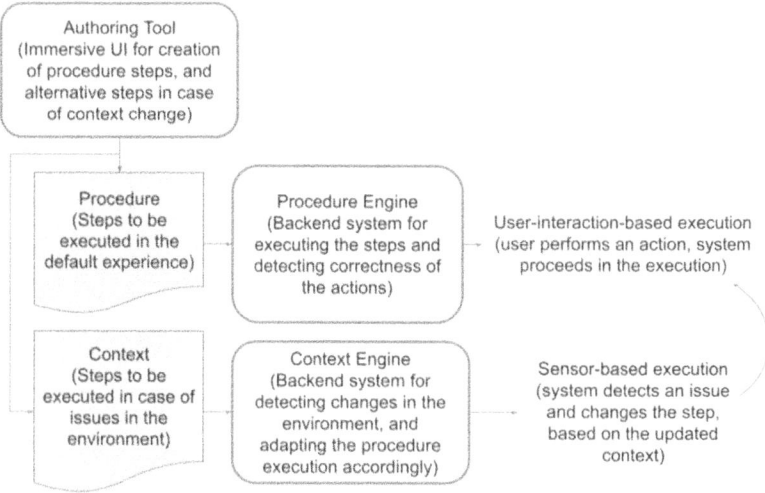

Fig. 1. The proposed system structure and components

3.2 Research Agenda

For the PhD research in general I have defined the following agenda:

- Initial identification of the problem;
- Research on the State of the Art, regarding End User Development and XR;
- Design, Prototyping and Implementation of usable tools for supporting end users in the authoring process;
- Validation of the result through user tests and/or comparisons with the current state of the art.

In particular, the work described in this paper incorporates the following additions to the general approach:

- The research on the State of The Art will also look for literature on vocational training educational content and how interactions are made in XR;
- The implementation of the tool will combine the knowledge from both XR and educational content to develop an initial interface, then design iterations will be made to refine the initial result and create a candidate for the user tests;
- Evaluation of the system will be initially done with an expert study, and subsequently with end users, in order to get opinions on the quality of the system and further improve it, to achieve a better result, following the suggestions of the participants. We will analyse the results of the evaluations and use them to iteratively improve the system. This process will be continued until the system is mature and meets the proposed goals.

4 Preliminary Results

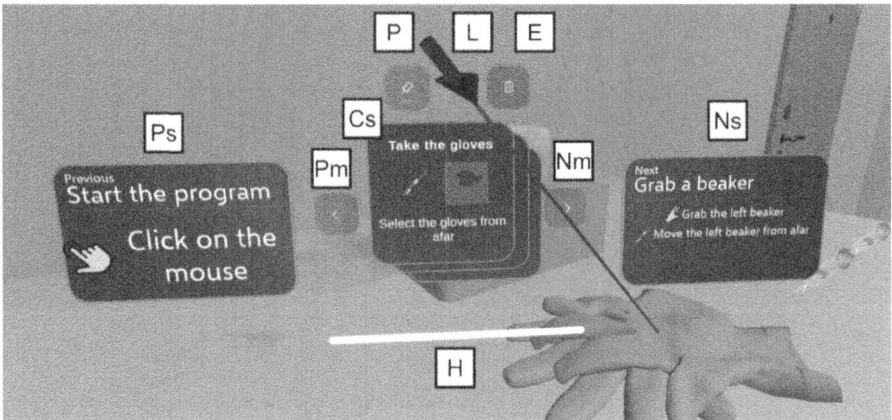

Fig. 2. Current Interface for the project. The middle panel (Cs) shows how to execute the current step, with the possibility to change modality, if multiple, with the arrow buttons (Pm and Nm). The left panel (Ps) shows the most recent successful step done, the right panel (Ns) shows the next step, with all the possible modalities. The Pin button (P) and the manipulation handle (H) allow movement of the interface, the Extended Info button (E) toggles visibility of the left and right panels. Tapping the middle panel shows the locator arrow and line (L) for identifying the object in the environment and teleporting the user near it, if applicable.

Research has been done on this project for identifying an initial state of the interface.

4.1 Workshop with Experts and Resulting Interface

A workshop was conducted with six HCI experts to gather insights and guidelines for designing an effective user interface for the system. After being introduced to the project's features, the experts were tasked with proposing a potential interface layout that adhered to the following criteria:

- Adaptability to XR devices;
- Adequate information presentation;
- Clear progress indication;
- Meaningful modality representation.

The suggestions of the experts formed the foundation for an initial functional version of the system. A notable recommendation that surfaced among all the experts was to implement a simple and minimal interface that could expand to reveal more information upon user demand, maintaining the current step in focus throughout the training process. This suggestion was incorporated into the current iteration of the interface, which consists of three panels (as seen in Fig. 2):

- Middle Panel (Fig. 2-Cs): always visible, displaying the current step and the available modalities via arrow buttons (Fig. 2-Pm and Fig. 2-Nm) in the panel's peripheries. The panel utilises a combination of text (step description and modality description) and icons (modality representation and target object indication) to guide the user's actions.
- Left Panel (Fig. 2-Ps): displays the latest successfully completed step, employing the same text descriptions and modality icons as the middle panel.
- Right Panel (Fig. 2-Ns): displays the next step to be undertaken, accompanied by a list of potential modalities.

Additional user-accessible functions include:

- Interface Movement: utilising either the automatic movement button (Fig. 2-P) or the manipulation handle (Fig. 2-H) for manual positioning.
- Target Object Location: upon tapping the middle panel, an arrow and a line indicate the location of the object within the virtual environment (Fig. 2-L). If the currently selected modality is "touch" or "grab" the user will be also teleported near the target object.
- Extended Interface Expansion/Collapse: the expansion button (Fig. 2-E) shows the previous and next step panels for a complete view, while pressing the same button again collapses the full interface, for a simple view.

This prototype implementation of the proposed system implements the Procedure Engine and Context Engine as described in Sect. 3.1, with some notable differences. First, the authoring tool is not yet available, as this project is still in a prototyping phase. Second, procedures are currently serialised JSON files, and the execution of one procedure or another is defined by their inclusion in a specific project folder. In future work, a user interface will be developed to handle this process without manual modification of project files. Another notable aspect of this prototype is that the procedure creator is responsible for determining the best modality to suggest depending on the environment, when creating the steps and their alternatives. Future work may explore the use of state-of-the-art information to make this decision automatically, offloading this task from the End User Developer.

4.2 Expert Study on the User Interface

An expert study has also been made, with the goal of assessing if the interface is able to convey information about what the user can do. Three experts have been invited (2 female & 1 male, and all of them in between 28 and 29 years old), each one having at least one year of working experience with Virtual Reality environments and are knowledgeable in the domain of Human Computer Interaction. The evaluation was structured in four parts:

Tutorial. In the first part the experts were introduced to the interface, explaining its goal and its various parts. Experts were able to ask questions about any doubts that they would have got during this phase.

Cognitive Walkthrough. In the second part, the experts performed three tasks, while having the possibility to provide feedback and criticism on the task itself, during and after the completion of the task, using a think-aloud method. After each task they have been additionally asked about the level of confidence while doing the task, using a 10-level scale. The three tasks, performed in an hypothetical chemistry lab scenario, were the following:

- Perform a task with multiple possible actions, with the suggested action as the target action to perform. The experts needed to perform a far interaction selection of a pair of gloves on a desk;
- Perform a task with a single action available. The experts were required to touch a mouse positioned on a desk adjacent to the desk where the gloves were placed.
- Perform a task with multiple possible actions, with another action as the target action to perform. The experts needed to perform a grab interaction on a beaker in the same desk, and bring the beaker to a nearby waste basket.

Heuristic Evaluation. In the third part they have been introduced them to the Eight golden rules by Ben Shneiderman [15] and they have been asked, in their opinion, how well applied were the rules with respect to the interface itself, using a 5-level Likert scale. After each rule the experts were able to comment and discuss further why they chose a particular mark. In this part the rule "Permit easy reversal of actions" was omitted, since it is not applicable to this system.

Semi-structured Interview. In the last part qualitative comments on the features of the system have been collected, any positive and negative comments about it, whether if they would have had ideas for improving the interface, and whether if they felt the level of feedforward implemented in the system would help an user in understanding how to do the task that the system is requiring him to do.

Results. A slightly lower confidence level was recorded for the first task (T1 = Mean: 7, Md: 8, SD: 2.16), while the subsequent tasks (T2 = Mean: 7.67, Md: 7, SD: 1.7; T3 = Mean: 7.67, Mo: 8, Md: 8, SD: 0.47) yielded improved confidence scores. This may be attributed to the lack of a warm-up period to acclimatise the experts to the system, leading to a lower confidence level at the outset, which subsequently increased with familiarity. Notably, one expert successfully completed T3 using a modality different from the one stipulated in the task description, demonstrating their understanding of the interface.

In the heuristic evaluation, the system performed well across the majority of the rules, with the exception of Rule 2 (Seek universal usability), which scored slightly lower than the others. A suggestion was made to avoid over-reliance on icons for conveying interaction and object descriptions, as non-expert users may find them challenging to comprehend. Instead, incorporating brief videos (i.e., direct feedforward) or an icon vocabulary help system could be beneficial.

Feedback on the Semi-structured Interview. The test conducted also highlighted some interesting comments on the interface. The experts found the interface to be easy to use, straightforward, and consistent. They particularly appreciated the fact that tasks were presented one at a time (P3), that all the information was gathered in one place (P1), and that the task flow was clearly explained (P2). They also felt that the system was consistent in presenting feedforward information (P2), and that it gave them a high level of control over their training (P2).

However, the experts also identified some limitations with the interface. They would have preferred a dedicated button for enabling object highlighting (P1, P2), or a better hint for clicking on the card stack for activating the feature (P2, P3) and they felt that there was a need for better explanations for how to perform actions (P1, P3). They also suggested that the system could provide more prominent visual cues for confirming the correct execution of an action (P2, P3), better interface clarity when showing the number of cards on only one alternative (P1) and the teleportation feature being available for every possible interaction, not only for touch and grab (P1).

4.3 Challenges and Future Works

In this section, I will discuss the main challenges identified so far, in both the general research and the latest project.

The first one concerns the research of a balance between what the user can do with a user interface of a given system, and what the system requires to be explicitly programmed through programming languages. Indeed, using End User Development focused systems give the user the possibility to define behaviour using a user-friendly interface, enabling it to create experiences that he could not have done without it. Notable interfaces that can be used, for example, are Natural-Language-supported interfaces (like ECARules4All [1]) that help the user define a behaviour by defining an human-readable sentence, or block-based programming (like Blockly [2]) that guides the user in the building of working code by abstracting the syntax using visual representationŝ of programming language elements. So, the more abstraction from code can be made, the less help from an expert developer can be required from a given system, to achieve the goal of the end user developer.

Moreover, another issue closely related to the previous one, is the idea that allowing too much abstraction in a given system, and by extension generalise it too much, could give the end user developer the added challenge to understand how to adapt the tool for his specific domain by learning how to program his idea into the environment, instead of programming it right away. These issues have been identified in the past, for example Yigitibas et al. [23] defined that modeling languages can be used to abstract the complexity of development, making it accessible to a wider range of people, including those without programming experience. However, learning a new modeling language can be a barrier to entry for novices and therefore, they believe that developing an interactive VR scene should require a minimal entry barrier to reach a maximum of novices. In future work, I will conduct research to find a balance between making the system general enough for a wide range of users and specific enough for end-user developers to define the actions that end users need to perform in real-world contexts.

A third element of concern can be borrowed from the field of the IoT. In this field, it may happen that using End User Development interfaces for programming the behaviour of an embedded system, they may bring some inconsistency between what the user wanted and what is the output of the system itself. In this case, the issue shows that there is the need of tools that help the end user understand how to solve the inconsistency, for example by implementing visual debuggers, like Corno et al. [5] proposed with "My IoT Puzzle". Indeed, this solution can be further explored and considered for XR environments as well, and I will provide insight on the matter in the next projects. End users may also need assistance in understanding the results of their interactions, especially when the system does not produce the expected outcome (e.g., when a user interacts with an object using a modality that the system does not currently require for moving forward in the procedure).

The final issue, specific to the latest project, may require a method for ensuring that, regardless of the content provided to the trainee, this is going to be well delivered to him. In this case, the system can be designed in a way that it becomes robust enough to support any kind of environment in general. Moreover, to assess the quality and the level of training of the learner it may still be a necessity to resort to the help of the trainer; in this case it is possible to relinquish him from some of these tasks, by distinguishing two different future versions of the system, with different outcomes: i) the system with a "non-empathised" type of interaction that can be used for autonomous evaluation. This version can be enough for learning the steps of the procedure, thus avoiding the evaluation of the trainer; ii) the system with a "precise" type of interaction. This version, instead, requires a path validation, and by extension a proper trainer evaluation (e.g. when a precise rotation of a screw, or a precise position of a piece, is required to complete the procedure in the optimal case). We will conduct further evaluations in the future, as the project progresses to more mature stages.

5 Conclusions

This paper presents a possible solution to the creation of XR experiences enhanced by End User Development principles, currently focused on vocational training. The methodology consists in the creation of a system that lets the user define steps for a training procedure through an authoring tool, and then be able to migrate said procedure throughout the XR spectrum. This system is also capable to adapt the procedure using the environment context of the user, through the use of sensors and user preferences. The described research is currently being developed during my visit to leading research groups in my period abroad. In the future, I will also collaborate with a leading ICT Italian company, whose work is mainly focused on the valorisation of the cultural and environmental heritage, for exploring other possibilities in the field of Extended Reality and End User Development.

6 University Doctoral Program and Context

The PhD project is led by Valentino Artizzu, a second-year PhD student in Computer Science in the Department of Mathematics and Computer Science of the University of Cagliari (Italy), starting from January 2022. The expected end of the program is in December 2024. He has a Master's Degree in Computer Science at the University of Cagliari in 2021, with a thesis based on the implementation of the rule engine that later became one of the main components of ECARules4All [1], after a scholarship and the beginning of the PhD program. His main research interests are XR technologies, Videogame Development and Design, and IoT. He works in the CG3HCI (Computer Graphics & Human Computer Interaction) research group, founded by prof. Riccardo Scateni, who leads the Computer Graphics part of the group, and later joined by prof. Lucio Davide Spano, who leads the Human Computer Interaction one.

Acknowledgments. This work is supported by the Italian Ministry and University and Research (MIUR) under the PON program: The National Operational Program "Research and Innovation" 2014–2020 (PON R&I).

References

1. Artizzu, V., et al.: Defining configurable virtual reality templates for end users. Proc. ACM Hum.-Comput. Interact. **6**, 163:1–163:35 (2022). https://doi.org/10.1145/3534517
2. Blockly. https://developers.google.com/blockly. Accessed 28 Aug 2023
3. Cassola, F., Pinto, M., Mendes, D., Morgado, L., Coelho, A., Paredes, H.: A novel tool for immersive authoring of experiential learning in virtual reality. In: 2021 IEEE Conference on Virtual Reality and 3D User Interfaces Abstracts and Workshops (VRW), Lisbon, Portugal, pp. 44–49 (2021). https://doi.org/10.1109/VRW52623.2021.00014.
4. Chiang, F.-K., Shang, X., Qiao, L.: Augmented reality in vocational training: a systematic review of research and applications. Comput. Hum. Behav. **129**, 107125 (2022). https://doi.org/10.1016/j.chb.2021.107125
5. Corno, F., De Russis, L., Monge Roffarello, A.: My IoT puzzle: debugging IF-THEN rules through the jigsaw metaphor. In: Malizia, A., Valtolina, S., Morch, A., Serrano, A., Stratton, A. (eds.) IS-EUD 2019. LNCS, vol. 11553, pp. 18–33. Springer, Cham (2019). https://doi.org/10.1007/978-3-030-24781-2_2
6. Djajadiningrat, T., Overbeeke, K., Wensveen, S.: But how, Donald, tell us how? On the creation of meaning in interaction design through feedforward and inherent feedback. In: Proceedings of the 4th Conference on Designing Interactive Systems: Processes, Practices, Methods, and Techniques, pp. 285–291. Association for Computing Machinery, New York (2002). https://doi.org/10.1145/778712.778752
7. Eroglu, S., et al.: Design and evaluation of a free-hand VR-based authoring environment for automated vehicle testing. In: IEEE Virtual Reality and 3D User Interfaces (VR), Lisbon, Portugal, pp. 1–10 (2021). https://doi.org/10.1109/VR50410.2021.00020
8. Fanni, F.A., et al.: PAC-PAC: end user development of immersive point and click games. In: Malizia, A., Valtolina, S., Morch, A., Serrano, A., Stratton, A. (eds.) IS-EUD 2019. LNCS, vol. 11553, pp. 225–229. Springer, Cham (2019). https://doi.org/10.1007/978-3-030-24781-2_20
9. Fungus Games: Fungus. https://fungusgames.com. Accessed 28 Aug 2023
10. Furht, B.: Virtual reality. In: Furht, B. (ed.) Encyclopedia of Multimedia, pp. 968–968. Springer, Boston (2008). https://doi.org/10.1007/978-0-387-78414-4_255
11. Garzotto, F., Gelsomini, M., Matarazzo, V., Messina, N., Occhiuto, D.: XOOM: an end-user development tool for web-based wearable immersive virtual tours. In: Cabot, J., De Virgilio, R., Torlone, R. (eds.) ICWE 2017. LNCS, vol. 10360, pp. 507–519. Springer, Cham (2017). https://doi.org/10.1007/978-3-319-60131-1_36
12. Green, T.R.G., Petre, M.: Usability analysis of visual programming environments: a 'cognitive dimensions' framework. J. Vis. Lang. Comput. **7**, 131–174 (1996). https://doi.org/10.1006/jvlc.1996.0009
13. Lieberman, H., Paternò, F., Klann, M., Wulf, V.: End-user development: an emerging paradigm. In: Lieberman, H., Paternò, F., Wulf, V. (eds.) End User Development, pp. 1–8. Springer, Dordrecht (2006). https://doi.org/10.1007/1-4020-5386-X_1
14. Sacks, R., Perlman, A., Barak, R.: Construction safety training using immersive virtual reality. Constr. Manag. Econ. **31**, 1005–1017 (2013). https://doi.org/10.1080/01446193.2013.828844
15. Shneiderman, B., et al.: Designing the User Interface: Strategies for Effective Human-Computer Interaction. Pearson, Boston (2017)

16. Stone, R.T., McLaurin, E., Zhong, P., Watts, K.: Full virtual reality vs. integrated virtual reality training in welding. Weld **92**, S167–S174 (2013)
17. UNESCO Institute for Statistics - International Standard Classification of Education ISCED 2011, p. 14. https://uis.unesco.org/sites/default/files/documents/international-standard-classification-of-education-isced-2011-en.pdf. Accessed 7 Nov 2023
18. Unity Real-Time Development Platform | 3D, 2D, VR & AR Engine. Unity. https://unity.com. Accessed 2 Sept 2023
19. Unreal Engine | The Most Powerful Real-Time 3D Creation Tool. Unreal Engine. https://www.unrealengine.com/en-US. Accessed 2 Sept 2023
20. Wolfartsberger, J., Niedermayr, D.: Authoring-by-doing: animating work instructions for industrial virtual reality learning environments. In: 2020 IEEE Conference on Virtual Reality and 3D User Interfaces Abstracts and Workshops (VRW), Atlanta, GA, USA, pp. 173–176 (2020). https://doi.org/10.1109/VRW50115.2020.00038
21. Wu, H.-Y., et al.: Designing guided user tasks in VR embodied experiences. Proc. ACM Hum.-Comput. Interact. **6**, 24, EICS, Article no. 158 (2022). https://doi.org/10.1145/3532208
22. Yigitbas, E., Klauke, J., Gottschalk, S., Engels, G.: VREUD - an end-user development tool to simplify the creation of interactive VR scenes. In: IEEE Symposium on Visual Languages and Human-Centric Computing (VL/HCC), St Louis, MO, USA, pp. 1–10 (2021). https://doi.org/10.1109/VL/HCC51201.2021.9576372
23. Yigitbas, E., et al.: End-user development for interactive web-based virtual reality scenes. J. Comput. Lang. **74**, 101187 (2023). https://doi.org/10.1016/j.cola.2022.101187

Exertion Trainer: Smartphone Exergame Design to Support Children's Kinesthetic Learning Through Playful Feedback

Carla Gómez-Monroy(✉) and Alejandro C. Ramírez-Reivich

Universidad Nacional Autónoma de México, Mexico City, Mexico
carlagm@comunidad.unam.mx, areivich@unam.mx

Abstract. Learning Apps have come a long way; nevertheless, Apps focused on physical education are underrepresented. This research explores exertion games that enable smartphone users to playfully learn the correct execution of target physical activities, as it is the straight punch in boxing. By implementing the sensors and actuators commonly found on modern smartphones, the proposed exertion game measures the movement of the user's punching hand while performing a straight punch. The Exertion Trainer uses the resulting smartphone accelerometer data to evaluate the users' techniques and provides sensory feedback to correct them. A graphic user interface was developed to convey the example movement that the user should replicate. A vital feature of this research is the user-centered design approach, which aims to generate new human-computer interactions by exploiting existing and pervasive mobile devices. So far, an expert user performance has been assessed to define a data baseline of the correct technique of the straight punch. Preliminary tests were done with two users. An A/B comparative experiment is proposed to validate this research. The control group, a traditional video-based tutorial, will be compared to the experimental group, a gamified interaction with the proposed exertion game. Participants will engage in a quantitative experiment (performance) and a qualitative questionnaire (experience). This study aims to transform the time users spend on the screen from visual content consumption to physical activation.

Keywords: Interaction Design · Human-Computer Interaction · Exertion Game · Learning through Smartphones · User-Centered Design · Movement-Based Design · Empirical studies in ubiquitous and mobile computing

1 Introduction

As smartphones have become a pervasive technology, the scope of available Apps has increased in complexity and variety, including Learning Apps tackling complex subjects such as math, music, and language. A fraction of such Learning Apps has attempted to increase user engagement through the gamification of the target discipline [1]. Nevertheless, learning games focused on Physical Education (PE) are underrepresented. Even though there are games for mobile devices that promote vigorous physical activity from the user, what Mueller et al. called exertion games [2], exertion games designed to improve the user's proficiency in a specific physical discipline or sport are rare, meaning that there is a gap. Our motivation is to implement the sensors and actuators on modern smartphones to assess the user's physical characteristics and experience in a given sport to provide personalized feedback in a playful App. In other words, our goal is to study exertion games as a learning tool for PE. We will compare the efficacy of a bespoke exertion game that provides directions to improve the user's technique in a target physical activity, a straight punch in boxing, as it is currently done in professional and high-performance training centers, where technology is used to trace and measure speed and acceleration on athletes. Ceballos and Rodriguez [3], in 2014, studied the straight punch in boxing to classify technique mistakes by analyzing the athlete's body movement through video and calculating the corresponding acceleration of the limps. Nevertheless, Ceballos and Rodriguez's approach analyzes the punch after the practice while we focus on the user's personalized feedback during the practice.

2 Related Work

There is current research in human motion focused on smartphone sensors [4, 5], user experiments [1, 6], feedback systems [7–9], technologies that encourage movement [10], a digital fabrication that senses dance [11], and exertion games [2, 12, 13]. Still, other commercial technologies relevant to this research are floor sensing systems and physical activation [14], physical activation systems on game consoles or personal computers [15, 16], mobile applications for fitness training that do not use smartphone sensors [17], and mobile boxing training applications with portable external sensing [18]. As well as research focused only on the straight punch in boxing [3, 19–21].

The World Health Organization reported, "It is clear that in too many schools in too many countries, there is a record of failure in physical education. Children are being denied the opportunities that will transform their lives." Schools in some countries offer an average of two hours per week of Physical Education (PE); in other cases, it is even lower or none. The instruction and infrastructure quality are the main challenges [22].

The COVID-19 pandemic highlighted educational challenges with online learning, such as technological access, pedagogical quality, isolation, and health [23, 24]. However, education prioritized math and reading skills over the already lagging physical education content, so many schools cut PE online classes. Many children spend enormous amounts of time sitting in front of a screen for schooling, movies, TV, video games, eating, and commuting. Ishihara et al. stated that "exercise during childhood is essential for brain development and [… to] have higher cognitive functions in later life" [25]. Furthermore, it is well established that exercise benefits the cardiovascular, motor, nervous, muscular, and skeletal systems, and general and specific psychomotor skills.

3 Research Question

This research aims to develop an engaging user interface (exertion game) that motivates users to learn and develop anywhere kinesthetic skills with the correct body technique by interacting playfully and unconventionally with smartphone sensors and actuators. For example, wear the smartphone on the leg to sense the user's leg acceleration while performing a cartwheel. Alternatively, wear the smartphone on the punching hand while performing a straight boxing punch; see Fig. 1. Thus, users transform their time on the screen from visual content consumption to physical activation.

Fig. 1. (A) and (B) show the Exertion Trainer animation screenshots, where (A) is the initial/final position or on-guard stance, and (B) represents the full extension of the straight punch. (C) and (D) show the colocation of the smartphone as a wearable device on the participant's wrist, secured by a commercial sports wrist phone holder, where (C) is the user replicating (A) 's, and (D) is the user throwing a straight punch like in (B).

The main research question is: **How do we design Apps** (using the smartphone as a non-conventional wearable) **to support children's kinesthetic and playful learning through interaction design**? The sub-questions are:

- How to teach the technique of a straight punch through a smartphone exertion game?
- How to track bodily motion using the smartphone as a wearable?
- How to provide personalized feedback when applying a smartphone as a wearable device?
- How to compare the acceleration curves of an Expert User with those of End Users (on the fly and in a stand-alone smartphone App)?

4 Approach

During the COVID-19 pandemic, we identified a gap in human-computer interaction (HCI). Even though Mexican children were taking their normal subjects (math, language, art) during online schooling, virtual classes do not provide a viable platform for teaching kinesthetic skills, not because it is hard to "give" instructions over a webcam, mobile app, or videogame. Instead, the problem lies in the evaluation method of each student. New ways of allowing teachers and researchers to continue their PE classes and propose new learning strategies that engage youngsters at a distance while preserving their privacy are needed.

Therefore, this research focuses on studying and exploring the feasibility of developing new human-computer interactions by exploiting existing and pervasive mobile devices. Moreover, we chose a hands-on strategy and a user-centered design methodology, where working closely and iteratively with the end user is the primary tool to generate knowledge. Each design cycle of the —Exertion Trainer— prototype, a tailored mobile exertion game to practice the straight punch, casts insights into the problem of applying modern smartphone sensors and actuators as wearable devices to track bodily motion and provide playful personalized feedback to children. We propose research validation through a user study, evaluating this human-computer interaction scenario's quantitative (data from smartphone sensors) and qualitative (user experience) aspects.

We selected the straight punch as a case study of physical activity because it is an extensively studied technique [3, 19, 20]. Moreover, the biomechanics of the movement can be described within a few degrees of freedom [3, 20]. The few degrees of freedom are to avoid problems reported by d'Ávila et al., who attempted to teach dancing through a wearable device that produced sensory feedback in the form of pressure on the user's limbs, aiming to direct the dancer's choreography. The authors concluded that even though their "Wearable Choreographer" influences the dancer, there are no apparent signs of a teaching/learning process [7].

Furthermore, we chose the smartphone platform due to its pervasive nature and scalability properties for research as well as for real-life application, as shown in the research "The Guardians: Designing a Game for Long-term Engagement with Mental Health Therapy" by Ferguson et al. in 2021, where the authors developed a mobile gaming application to promote the acquisition of healthy behaviors in users. This application is based on mental health therapy. In the early stages of development, they tested the application with focus groups of 14 individuals. Afterward, they published it on PlayStore and AppStore, obtaining and analyzing game interaction data from 7,782 real users [1].

Lastly, the technical question, is it possible to assess the user pose with the smartphone as a wearable device? It was tackled by Ahuja et al. in 2021 in their work "Pose-on-the-Go: Approximating User Pose with Smartphone Sensor Fusion and Inverse Kinematics." The researchers estimate the pose of the entire human body through data acquired by sensors incorporated in a smartphone. Nevertheless, inverse kinematics processed the data in an external computer [4].

5 Research Methodology

The research methodology is divided into two blocks. First, design an interactive App for teaching children kinesthetic skills (Exertion Trainer), and second, evaluate such prototype through a user study.

For the design of the Exertion Trainer, we applied the six user-centered design principles found in the ISO 9241–210:2019 *"Ergonomics of human-system interaction - 210: Human-centred design for interactive systems"* [26], as well as the five principles for designing exertion games published by Mueller et al. [2]. Concerning the in-app data analysis strategy to evaluate, on the fly, the user's technique after each punch, we applied the performance-expert approach by Ericsson (2015) [6]. In the Exertion Trainer case, an expert user was asked to perform several repetitions of a straight punch to characterize the acceleration curves of the correct technique as the baseline. Afterward, the baseline was embedded in the Exertion Trainer code to be compared to the acceleration curves of the user study participants. Even when the system is trained to recognize in real-time if the user is performing correctly or incorrectly, such experimental data will be stored for further analysis by the researcher and the expert user when the experiment is concluded.

To evaluate the Exertion Trainer, the participants will engage in a quantitative experiment (performance) and a qualitative questionnaire (experience). Thus, we designed an A/B comparative experiment with 30 participants, 15 in group A and 15 in group B. The control group (A) corresponds to a traditional interaction (video-based tutorial with no feedback to the user in real-time), while the experimental group (B) corresponds to the gamified interaction with the proposed exertion game (the Exertion Trainer that includes feedback in real-time). Each participant will have 20 min of practice with their assigned learning technology (video-based tutorial or the Exertion Trainer) and will be asked to answer a complementary written user experience questionnaire developed based on the Laugwitz, et al. guidelines [27]. The quantitative evaluation (performance) is two-folded and will take place afterward by the researcher (data-centric) and by the expert user (video-based). The results for both groups will classify each participant's technique improvement with a seven-point Likert scale. The researcher will analyze participants' smartphone accelerometer data logged while wearing the device and executing the straight punch sequence. The expert user will rate the participants' videos taped during the user study. The outcomes of the user experience questionnaire, the researcher data Likert scale, and the expert user video Likert scale will be compared.

During the user study, the Exertion Trainer will only provide feedback to experimental group participants. The user will receive one instruction. The system will average the data collection in a given amount of time, then the feedback system will be triggered through text-to-speech and sound effects, and the user will have enough time to do as recommended. The system tags the data with those two moments: 1) after instruction and 2) after feedback. The provided amount of time will serve to calibrate the system at the beginning; the time will be shortened throughout the repetition of the sequence. The feedback in real-time will be prioritized based on the error classification to achieve the baseline. Feedback prioritization is required because, firstly, the system is not as fast as giving text-to-speech feedback as it is to log the data. Secondly, the user may need to correct more than one position per movement and get confused if multiple recommendations are provided almost simultaneously (Fig. 2).

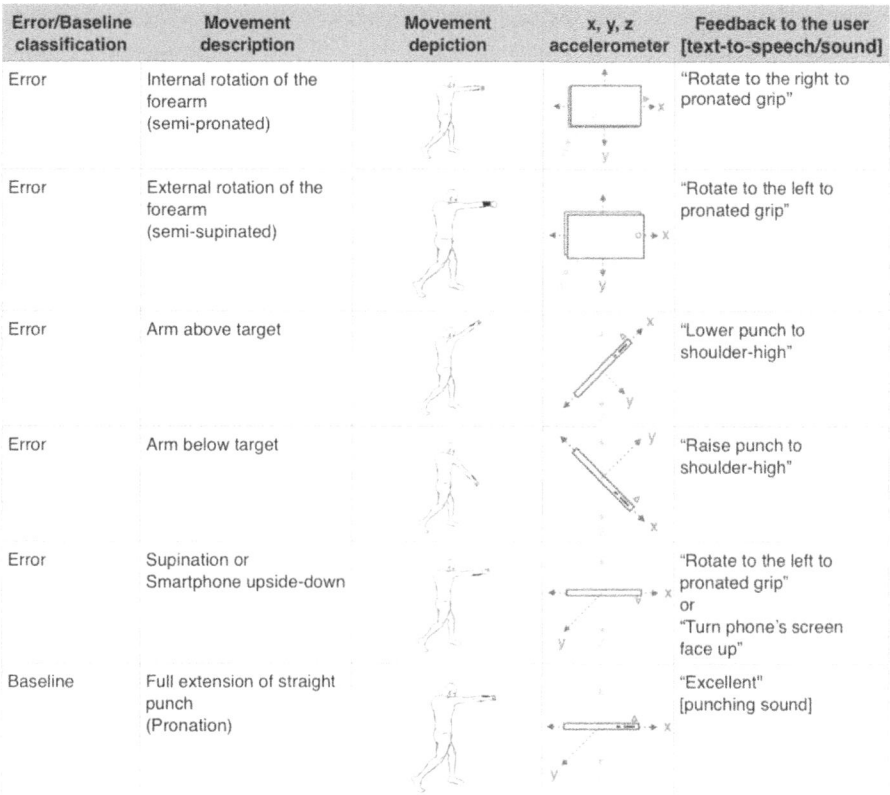

Fig. 2. Exertion Trainer (experiment group) feedback prioritization table for the full extension of the straight punch.

6 Results and Discussion

We have designed and coded a functional mobile App for Android smartphones that includes a bespoke, gamified graphic user interface, see Fig. 1, a data acquisition system that logs the smartphone sensors' data in the device as well as sends and stores the data into Firebase. Furthermore, an expert user performance has been assessed to define a data baseline of the correct technique of the straight punch. Thus, the system detects the child's errors in real-time and automatically triggers predefined verbal recommendations. Preliminary tests with human-in-the-loop design of sensing system, user interface, human-computer interaction, and preferred feedback types were done with two users. The expert user acceleration curves are more consistent across events and show greater acceleration peaks than the preliminary tests. Therefore, an A/B comparative experiment with 30 children is proposed to validate this research.

Ferguson et al. found that players must feel connected to the game [1]. Therefore, the gamified graphic user interface was developed to convey the precise example movement through animated cartoons that the user should replicate, see Fig. 1. Also, the standardized tasks of the straight punch were studied and sketched in a sample animation for the users.

In the preliminary tests, the two users were crucial for selecting the sensing system for the Exertion Trainer. One of the three tested systems was the PoseNet model (similar to Google Pose Detection). PoseNet detects in real-time 17 human body points by the pose estimation model of the person in front—in this case—of the smartphone's camera [28]. As illustrated in Fig. 3, when the target user performs a straight punch facing the smartphone's camera, PoseNet's 2D system is not as accurate as if the user's body is sideways to the camera. However, the user turns his head towards the screen, making the learning technically incorrect. When the body is sideways to the camera, the data recorded is correct; however, the visual feedback is not always precise, confusing the user. Five other factors are unsuitable for this research: one is that the smartphone has to be placed away from the target user (in countries like Mexico, using the Exertion Trainer in public places may add a risk of getting it stolen while interacting with it). The second factor is that children with visual refractive errors tend to get closer to the screen,

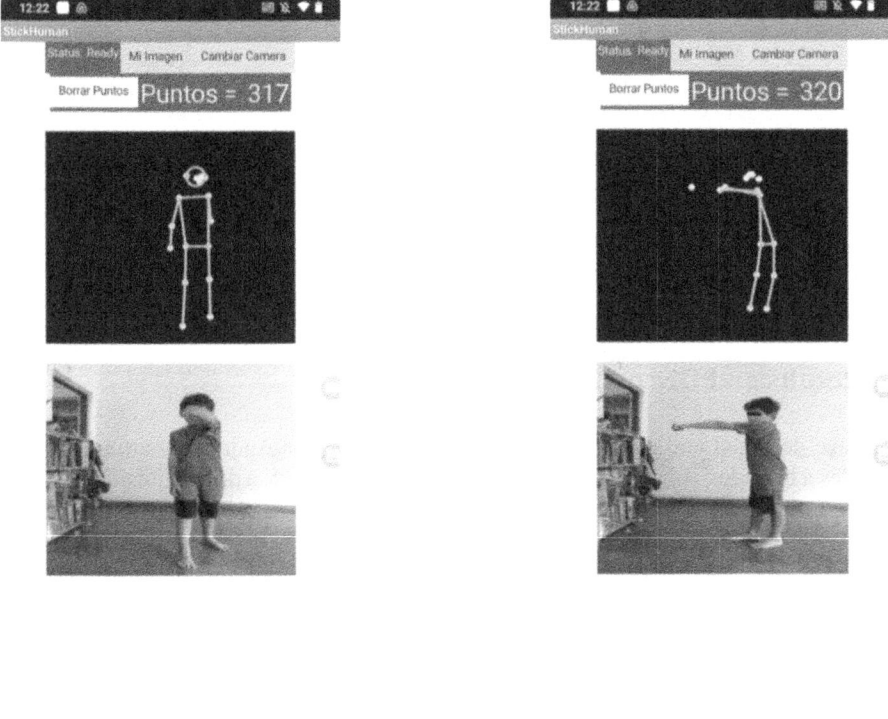

Fig. 3. Preliminary test with PoseNet. The left image shows the target user performing a straight punch in front of the smartphone. The right image depicts the user sideways to the smartphone's camera.

having their extremities out of the measurements. Then, the user will move back and forth, returning imprecise experimental data. Third, it is impossible to have the PoseNet running on the backend since the mobile phone is attached to the body, and the camera will face the ceiling, the floor, or other users around the studied user. The fourth reason why PoseNet was not chosen for this research is that children look at the screen to get visual feedback, and one goal of this research is to reduce the visual interaction with the smartphone and encourage children to use other senses when interacting with the device. Finally, the fifth factor is that some parents are concerned about their children's privacy and safety by having the image of their children recorded if the camera is activated, regardless of whether PoseNet runs in the front or back end and only stores keypoints.

The second sensing system tested was the accelerometer MPU6050 with an Arduino Nano, and the third system was the Exertion Trainer installed in the Motorola Moto G30. The resulting acceleration curves of the straight punch tested by the same target user were similar. Thus, the researcher chose to run all the user trials with the same Motorola Moto G30 device (See Sect. 7 Considerations for Designers).

As shown in Fig. 4, we obtained the Exertion Trainer straight punch acceleration graphs similar to those from other researchers, such as Kimm and Thiel [20], who used an accelerometer inside a boxing glove, and Ceballos and Rodriguez [3], who used a post hoc video analysis strategy. The above validates the experimental model since the experimental data obtained with the Exertion Trainer prototype have the sensitivity and fidelity necessary to study the straight punch established in the literature.

The Exertion Trainer code analyzes the four sections of the straight punch and helps identify errors in the straight punch technique. For example, dropping the fist during the punch is an error. Also, not fully extending the arm contradicts the correct technique. Nevertheless, classifying this and other technique errors must be improved in the exertion game.

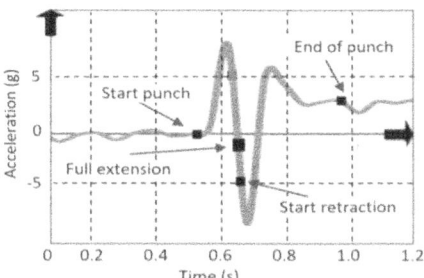

Fig. 4. Experimental data of the prototype: acceleration graph of a straight punch obtained with the Exertion Trainer game. Within the curve, the four straight punch tasks as defined by Kimm and Thiel [20] are highlighted: 1) the start punch, 2) the full extension, 3) the start of retraction, and 4) the end of the punch.

7 Considerations for Designers

We recognized the enormous challenge of developing mobile Apps (using the smartphone as a non-conventional wearable) that asses the user's body movements in real-time. Yet, we deem it possible to achieve it. Consequently, we sum up some considerations for interaction designers in this work.

Even though accelerometers have been embedded into smartphones for more than fifteen years, we have found tremendous reluctance from experts in the HCI field to use smartphones as reliable motion capture systems for this research. However, the results of Grouios, et al. "showed that there were not significant differences between the three current generation smartphone[s] and the Vicon MX motion capture system mean acceleration data" [29]. They compared the weaknesses and strengths of the studied motion capture systems in a table, listing more strengths and fewer weaknesses for the tested smartphones than for the Vicon MX motion capture system [29].

Following, we share some other design considerations.

Unique to the Exertion Trainer Methodology

- The personalized feedback is culturally dependent and language-specific.
- The exertion baseline has to be obtained per each skill to be taught.
- Each skill has to be subdivided into measurable tasks.
- Defining the correct technique of a skill requires expert user involvement.
- Experts define the feedback framework.
- The feedback has to be engaging and ensure user understanding.
- The design must consider anatomical differences and individual basic physical qualities.
- The continuous and close collaboration of children (target end-user), parents, and at least an expert user is required as part of the human-centered design methodology through the design of the data acquisition system, feedback system, data baseline, and graphical user interface.
- The human motion capture and user feedback systems are a single-gamified downloadable App for Android smartphones used as wearables.

Technology-Wise

- Smartphones (used as wearables) have no framework for human body biomechanical motion tracking standards, let alone for developing interacting kinesthetic Apps for children.
- Accelerometer inherent bias, drift, and noise require filtering.
- A non-inertial reference frame—the smartphone accelerometer—makes collecting precise measurements challenging.
- Using the smartphone as a wearable provides accelerometer data of a specific body part.

8 Conclusion

There is plenty of room to continue investigating **how we design Apps** (using smartphones as a non-conventional wearable) **to support children's kinesthetic and playful learning through interaction design**. Nevertheless, this research's lessons learned are twofold, technology-wise and unique to the Exertion Trainer methodology. The former relates to the usage of smartphones as wearables, given that the accelerometer data is from a specific body part based on a non-inertial reference frame and requires filtering of its inherent bias, drift, and noise. Besides, there is a lack of framework for human body biomechanical motion tracking standards helpful to develop interactive Apps for children. The latter—the Exertion Trainer methodology—includes a human-centered design that engages children as target users (and their parents), interactive instructional feedback, body-tracking, real-time data acquisition, and a gamified graphical user interface. In addition, the Exertion Trainer methodology considers anatomical differences and age-specific language for instructional feedback. The Exertion Trainer is a single-gamified downloadable App for Android that turns smartphones into a human motion capture wearable device.

Interaction design was fundamental to obtain the baselines and error classification from the expert user's performance of a straight boxing punch. Also, the expert user provided the exertion baselines to teach each skill and its subdivision into measurable tasks, error classification, and the feedback framework. Moreover, preliminary tests with two users revealed key factors for choosing parameters for the Exertion Trainer sensing and feedback systems.

On the one hand, when comparing the Exertion Trainer App against a camera-based tracking system, the results showed that the camera for detecting human pose disrupted the user's attention. On the other hand, when comparing the Exertion Trainer App, running on a Motorola Moto G30, versus an MPU6050 accelerometer connected to an Arduino Nano (both as wearable devices), the data demonstrated that the acceleration curves were highly similar. Therefore, using a smartphone as a wearable to track body motion has proven successful.

9 Future Work

The future work and expected outcomes of this research are to

1. Conduct the proposed A/B comparative experiment.
2. Assess more than one body part: the smartphone can be placed in different body places and have the corresponding challenges within the App (for example, first focus on learning the correct right arm technique of the straight punch, then the proper left arm technique).
3. Explore if one user can wear more than one smartphone on different body parts by using other family members' or friends' phones and obtaining data similar to the Sony Mocopi 3D Motion Capture System.
4. Develop collaborative and dynamic kinesthetic teaching-learning interactions among users.
5. Ignite the creation of a database of human movements and routines that can feed artificial intelligence systems for physical education, motor skill development, learning sports, and other kinesthetic activities,
6. Extend our understanding of designing smartphone technologies as wearables for children's kinesthetic and playful learning and development,

7. Contribute to a mobile exertion game that supports the learning of children's bodily movements in real-time,
8. Define a set of HCI design and evaluation guidelines for further researching children's kinesthetic learning,
9. Evaluate if the Exertion Trainer could be instrumental in sports science and rehabilitation of motor skill limitations.

The expected social impact of this research is to reduce sedentary lifestyle-related problems like obesity, diabetes, and depression.

Acknowledgements. Many thanks to Vicente Borja, Víctor González-Sánchez, José Luis Jiménez-Corona, and Adriana Lira-Oliver for sharing their expertise through their helpful suggestions.

We are deeply grateful to boxing coaches Raúl Rocha and Raúl Rocha Jr. For tirelessly sharing their know-how and being the expert users. Likewise, we acknowledge the volunteers for the preliminary tests.

We appreciate the graphics and animations done by D. Daniela Zavala Montero and Axel Gael Olmedo Corona. Also, thanks to Agustín Martínez Figueroa for contributing to the Graphical User Interface code.

The first author is obliged to Consejo Nacional de Humanidades, Ciencias y Tecnologías (CONAHCYT) for the fully-funded Ph.D. fellowship. In addition, this research was funded by Universidad Nacional Autónoma de México through Centro de Diseño Mecánico e Innovación Tecnológica (CDMIT), Programa de Apoyo a los Estudios de Posgrado (PAEP) 2023, and for the technological equipment to validate this research: grant number PAPIIT - IT102122.

References

1. Ferguson, C., Lewis, R., Wilks, C., Picard, R.: The guardians: designing a game for long-term engagement with mental health therapy. In: IEEE Conference on Games (CoG), pp. 1–8. IEEE, Copenhagen (2021). https://doi.org/10.1109/CoG52621.2021.9619026
2. Mueller, F., Khot, R.A., Gerling, K., Mandryk, R.: Exertion games. In: Foundations and Trends® in Human–Computer Interaction, vol. 10, no. 1, pp. 1–86. Now Publishers Inc. United States (2016). https://doi.org/10.1561/1100000041
3. Ceballos Rubio, M.A., Rodríguez Rodríguez, M.: Análisis cinemático del golpe recto a la cara a larga distancia, de tres boxeadores de la categoría 13–14 años de la EIDE provincial Guantánamo [Kinematic analysis of the straight punch to the face at long distance, of three boxers of the 13–14 years old category of the EIDE Guantánamo province]. In: EFDeportes.com, Revista Digital. Buenos Aires, vol. 19, no. 193, June 2014. EFDeportes Homepage. https://www.efdeportes.com/efd193/analisis-cinematico-del-golpe-recto-de-boxeadores.htm
4. Ahuja, K., Mayer, S., Goel, M., Harrison, C.: Pose-on-the-go: approximating user pose with smartphone sensor fusion and inverse kinematics. In: Proceedings of the 2021 CHI Conference on Human Factors in Computing Systems (CHI '21), Article 9, pp. 1–12. Association for Computing Machinery, New York (2021). https://doi.org/10.1145/3411764.3445582
5. Oyebode, O., Ganesh, A., Orji, R.: TreeCare: development and evaluation of a persuasive mobile game for promoting physical activity. In: 2021 IEEE Conference on Games (CoG), pp. 1–8. IEEE, Copenhagen (2021)
6. Ericsson, K.A.: Acquisition and maintenance of medical expertise: a perspective from the expert-performance approach with deliberate practice. Acad. Med. **90**(11), 1471–1486 (2015). https://doi.org/10.1097/ACM.0000000000000939

7. d'Ávila Silveira, C.A., Afsar, O.K., Alaoui, S.F.: Wearable choreographer: designing soft-robotics for dance practice. In: Designing Interactive Systems Conference (DIS '22). pp 1581–1596. Association for Computing Machinery, New York (2012). https://doi.org/10.1145/3532106.3533499
8. Hülsmann, F., Göpfert, J.P., Hammer, B., Kopp, S., Botsch, M.: Classification of motor errors to provide real-time feedback for sports coaching in virtual reality — A case study in squats and Tai Chi pushes. Comput. Graph. **76**, 47–59 (2018). https://doi.org/10.1016/j.cag.2018.08.003
9. Woźniak, P., Knaving, K., Björk, S., Fjeld, M.: RUFUS: remote supporter feedback for long-distance runners. In: Proceedings of the 17th International Conference on Human-Computer Interaction with Mobile Devices and Services, pp. 115–124. ACM (2015). https://doi.org/10.1145/2785830.2785893
10. Saaty, M., Haqq, D., Toms, D. B., Eltahir, I., McCrickard, D.S.: A study on Pokémon GO: exploring the potential of location-based mobile exergames in connecting players with nature. In: Extended Abstracts of the 2021 Annual Symposium on Computer-Human Interaction in Play (CHI PLAY '21), pp. 128–132. Association for Computing Machinery, New York (2021). https://doi.org/10.1145/3450337.3483481
11. Wicaksono, I., Haddad, D.D., Paradiso, J.: Tapis Magique: machine-knitted electronic textile carpet for interactive choreomusical performance and immersive environments. In: Creativity and Cognition (C&C '22). pp. 262–274. Association for Computing Machinery, New York (2022). https://doi.org/10.1145/3527927.3531451
12. Exertion Game Lab's Projects page. https://exertiongameslab.org/projects. Accessed 13 Mar 2023
13. Yılmazer, H., Mueller, F., Coşkun, A.: Designing for bouldering: supporting kinesthetic learning of action sport enthusiasts. In: Designing Interactive Systems Conference (DIS '22 Companion), pp. 48–51. Association for Computing Machinery, New York (2022). https://doi.org/10.1145/3532107.3532884
14. Dance dance revolution. Wikipedia. https://en.wikipedia.org/wiki/Dance_Dance_Revolution. Accessed 03 June 2022
15. Just dance. Wikipedia. https://en.wikipedia.org/wiki/Just_Dance_(video_game_series). Accessed 05 June 2022
16. Jurkovich, T.: Fitness video games that will actually make you sweat. The Gamer. https://www.thegamer.com/fitness-video-games-active-exercise-gaming/. Accessed 03 Nov 2022
17. Links, R.: The 7 best personal trainer apps. In: Healthline. https://www.healthline.com/nutrition/personal-trainer-app. Accessed 10 Feb 2023
18. Johnny N.: Boxing punch trackers review 2022. In: Expert Boxing. https://expertboxing.com/boxing-punch-trackers-review. Accessed 06 Jan 2023
19. Sanchez-Rodríguez, D.A., Bohórquez-Aldana, A.F.: Análisis de la velocidad y la aceleración entre un golpe de boxeo y uno de taekwondo [Analysis of the speed and acceleration between a boxing and a Taekwondo punch]. In: Revista U.D.C.A Actualidad & Divulgación Científica, vol. 23, no. 1, Bogotá, Colombia (2020). https://doi.org/10.31910/rudca.v23.n1.2020.1481
20. Kimm, D., Thiel, D.: Hand speed measurements in boxing. In: 7th Asia-Pacific Congress on Sports Technology, APCST 2015. Procedia Engineering, vol. 112, pp. 502–506, ISSN 1877-7058. ScienceDirect (2015). https://doi.org/10.1016/j.proeng.2015.07.232
21. Merlo, R., García Sánchez, F.: Aspectos Relevantes sobre el Recto de Boxeo [Highlights of the Boxing Straight Punch]. In: G-SE, https://g-se.com/aspectos-relevantes-sobre-el-recto-de-boxeo-bp-d57cfb26ddaaf5. Accessed 10 Dec 2022
22. Hardman, K.: An up-date on the status of physical education in schools worldwide: technical report for the World Health Organisation. Published by the International Council of Sport Science and Physical Education (ICSSPE). Report from Berlin Physical Education World

Summit in November 1999 until 2007 updates. ICSSPE. https://www.icsspe.org/sites/default/files/Kenneth%20Hardman%20update%20on%20physical%20education%20in%20schools%20worldwide.pdf. Accessed 09 Oct 2022
23. Organisation for Economic Co-operation and Development: "Education at a Glance 2020: OECD Indicators." OECD Publishing, Paris (2020)
24. Organisation for Economic Co-operation and Development: "The State of School Education: One Year into the COVID Pandemic." OECD Publishing, Paris (2021). https://doi.org/10.1787/201dde84-en
25. Ishihara, T., et al.: Childhood exercise predicts response inhibition in later life via changes in brain connectivity and structure. In: NeuroImage, vol. 237, p. 118196. ScienceDirect, (2021). https://doi.org/10.1016/j.neuroimage.2021.118196
26. International Organization for Standardization: ISO 21001:2018 Educational organizations — Management systems for educational organizations — Requirements with guidance for use (2018)
27. Laugwitz, B., Held, T., Schrepp, M.: Construction and evaluation of a user experience questionnaire. In: Holzinger, A. (ed.) USAB 2008. LNCS, vol. 5298, pp. 63–76. Springer, Heidelberg (2008). https://doi.org/10.1007/978-3-540-89350-9_6
28. GitHub Homepage. https://github.com/tensorflow/tfjs-models/blob/master/pose-detection/README.md. Accessed 06 Sept 2023
29. Grouios, G., Ziagkas, E., Loukovitis, A., Chatzinikolaou, K., Koidou, E.: Accelerometers in our pocket: does smartphone accelerometer technology provide accurate data? Sensors **23**(1), 192 (2023). https://doi.org/10.3390/s23010192

Explaining Temporal Logic Model Checking Counterexamples Through the Use of Structured Natural Language

Ezequiel José Veloso Ferreira Moreira[✉] and José Creissac Campos

Departamento de Informática, Universidade do Minho and HASLab/INESC TEC, Braga, Portugal
`id9620@uminho.pt, jose.campos@di.uminho.pt`

Abstract. The use of model checking tools allows for the formal verification of properties over models of systems, improving their robustness. However, these tools are challenging to use, and their results require much work of interpretation to communicate to stakeholders. To address this issue, the IVY Workbench offers a plethora of options to make the process of creating and understanding the models, properties and results of the verification process more accessible, with a particular focus on interactive computing systems. Despite this, there is still a significant requirement of expertise to use the tool. To solve this, an approach to provide structured natural language explanations for the results of model checking-based tools is being developed, to be later incorporated into the IVY Workbench. This paper presents the current state of the approach's development, stating its objective and what results can already be achieved.

Keywords: Natural Language Generation · Model Checking · IVY Workbench · NuSMV

1 Introduction

Software has grown immensely over the last decades, becoming nearly omnipresent. As the field expanded, the need to ensure the safety and robustness of software has grown, something that can be achieved by applying formal methods tools and techniques [35]. Formal methods focus on the analysis, development and verification of computing systems through the use of mathematics to produce higher quality software [24].

Within formal methods, model checking is used to automatically verify formal properties over the behaviour of a formal model of a system [16], which in turn contributes to ensuring the finalised system's quality. However, the models used, and the results produced, by model checkers are formal in nature, requiring model checking experts to use them and translate their results into a form that the systems' domain experts can understand. This leads to difficulty in communication and hinders the adoption of these methods [4].

In the context of interactive computing systems, a number of tools to support formal modelling and verification have been proposed [5,10]. The IVY Workbench is one such tool [9]. To simplify the process of verification results analysis, the IVY Workbench provides features such as a tabular representation of the counterexamples produced during the verification process (see Fig. 1), as well as a state-by-state simulator that allows exploration of the model (see Fig. 2).

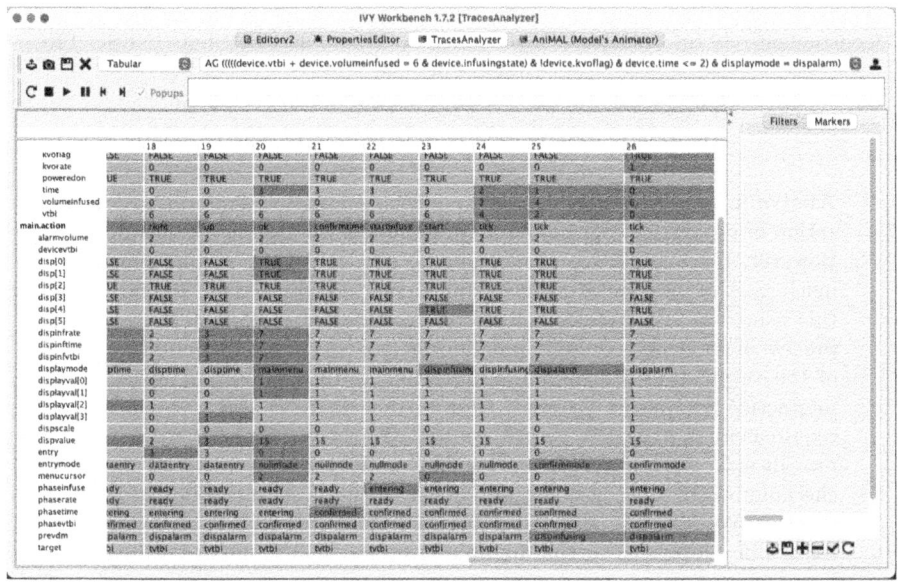

Fig. 1. IVY Workbench's tabular counterexample trace representation (a trace represents a behaviour of the system – columns correspond to states in the trace, lines represent attributes of the system, changes to the attributes are highlighted in yellow) (Color figure online)

Despite this, experience with the tool (cf. [11,23]) shows that substantial work is still required to translate the results produced by the tool into a form that non-experts can understand. Approaches are needed to support expressing the analysis results in the domain experts' language.

Considering the above, this paper aims to answer the following question: is it possible to define an approach to generate natural language explanations of the counterexamples produced by a model checker, so that they might be understood by domain experts who are not fluent in formal methods?

The paper describes the approach that is being taken to answer this question and is structured as follows: in Sects. 2 and 3, background and related work will be introduced; next, in Sect. 4, an explanation of the algorithm that is being developed will be provided, with an example of an explanation created by said algorithm presented in Sect. 5; the paper concludes with a discussion of future work in Sect. 6.

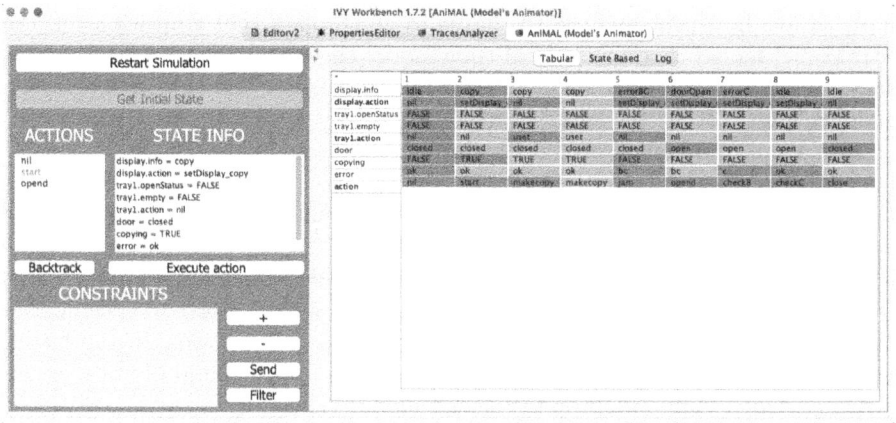

Fig. 2. IVY Workbench's state-by-state model simulator

2 Background

To better understand the problem this project aims to address, two areas of knowledge are relevant: Model Checking and Natural Language Generation. This section provides brief introductions to each of these areas.

2.1 Model Checking

Model checking focuses on the formal verification of a model's behaviour to determine if said behaviour obeys a series of properties. It is part of the formal methods area, which focuses on the specification, development and verification of computing systems through rigorous mathematical methods [35].

The work described herein focuses on temporal logic model checking, i.e., verification of temporal logic formulas over a state machine. Model checking can be formalised as follows [16]:

Theorem 1. *Let M be a Kripke structure (i.e., state-transition graph). Let f be a temporal logic formula (i.e., the specification). Find all states s of M such that $M, s \models f$.*

According to this definition, the model for a given system is a state-transition graph over which temporal logic formulas will be verified to determine if the model's behaviour obeys the temporal logic formulas. From this point forward, this document will always call the Kripke structure a model, and each temporal logic formula verified over it a property. These properties can be written in many different types of temporal logic, with LTL [32] and CTL [14] being common choices within the area.

Note that there exist variants of model checking, such as probabilistic model checking, where the transitions between states have an attached probability that dictates how probable the system is to go through that transition (cf. the PRISM [27] model checker), or real-time model checking, which focuses on verification of properties that include real-time as a factor (cf. the UPPAAL [3] model checker).

An intrinsic problem of model checking is state explosion [8]. That is, the model becoming too large for the analysis to be feasible. Tools such as NuSMV [12] address this by using symbolic techniques [8] and thus avoiding the explicit construction of the entire state-transition graph. Alternatives include the use of SMT solvers (cf. NuXMV [6]) or bounded model checking (cf. the CBMC [15] model checker). In the former case the solver resorts to specific theories (e.g. for natural numbers) to increase the reasoning power. In the latter, verification of a property only occurs until a given upper bound of state exploration, at which point the property is considered valid

For the type of model checking being considered in this work, tools include NuSMV and Spin [25], used in various real-world applications such as proving the correction of communication protocols [19] and railway systems [13].

One of the major advantages of using model checking tools is the ability to produce counterexample traces if a given property is violated by the model. This ability is fundamental in debugging complex systems [16]. The human intervention required during the process of using these tools can thus be divided into two parts:

– The first part of the process involves transforming the project requirements (usually written by stakeholders in natural language) into a formal model and attached properties.
– The second part requires analysis of the counterexamples produced by the verification process, typically by an expert in model checking, to understand the problems and devise effective solutions; stakeholders can play a relevant role in this analysis, but there is a need to express the counterexamples in a language that can be effectively communicated to them.

The explanation approach being developed aims to help with the second part of this process. The goal is to generate explanations for the results produced by model checkers such that understanding them does not require additional knowledge beyond that of the domain, leading to a simplification of the process of analysis, interpretation and communication of such results.

2.2 Natural Language Generation

With the above in mind, we can introduce the second area involved in the work: natural language generation. As the name implies, this area focuses on the generation of natural language (structured or not) as the output of a program [33]. It is part of the natural language processing area, which focuses on the interaction between humans and computing systems through the use of natural language [1].

The methods that exist to generate natural language text vary immensely, but a number of common tasks are done to create the final text [20]:

- **Content determination** - deciding which information should be communicated in the generated text.
- **Text structuring** - determining how the selected content must be structured in the generated text.
- **Aggregation** - structuring and ordering the information as meaningful sentences.
- **Lexicalization** - determining which words, terms and concepts to use to convey the text's information.
- **Referring expression generation** - determining how domain entities must be referred to within the generated text.
- **Linguistic realisation** - combining all the information generated in the other steps and organising/presenting it as a well-formed text.

While the tasks may be common, the ways and order in which they are done vary greatly among different architectures, of which some examples are [31]:

- **Pipeline architecture** - information flows sequentially among various well-defined components, each responsible for one or more tasks.
- **Revision architecture** - similar to the pipeline architecture, but with the added task of reviewing the text as it is generated within the components, altering it if necessary to produce better results.
- **Uniform architecture** - a single reasoning engine is responsible for dealing with the entire task of generating text.
- **Software design like architecture** - the generation of text is done by following software architectural model's principles

3 Counterexample Explanation Methods

Existing methods for generating counterexample explanations[1] vary significantly in how the explanations are generated and presented. However, many of the methods found can be classified into one of three different types of explanation: the simplification of the counterexample trace, the determination of probable cause for the problem that led to the violation of the property, the explanation of the property that was violated.

3.1 Simplification of Counterexample Traces

These explanation generation methods focus on generating and/or presenting a simplified version of the counterexample traces, i.e., focus on reducing the traces' interpretation difficulty.

---

[1] It should be noted that as far as this work is concerned, any method that reduces the requirements for understanding counterexamples is considered an explanation. This is because an explanation aims to better the understanding of something.

Examples of these types of explanation generation methods include:

- The minimisation of Bounded Model Checking counterexample traces proposed by Groce and Kroening [21], where two distinct algorithms are presented and compared with the purpose of generating a minimal counterexample trace that still contains the information required to understand how the property was violated within the counterexample.
- The generation of a visual explanation for **LTL** properties by Ovsiannikova et al. [30], where the trace is graphically represented, and relevant parts are highlighted. This is done by taking a **NuSMV** model, a **LTL** property and a counterexample, then encoding the model as modular blocks and presenting the counterexample by highlighting within these blocks the path taken in the counterexample and how the property was violated, as well as highlighting the causes of a given variable's value within each block.
- The use of graphical representations (e.g. tabular or activity diagrams) to represent a counterexample, as done by the IVY Workbench [9] and illustrated in Fig. 1. Additionally, the tool allows filtering or visually highlighting parts of the traces' representation according to given conditions, as shown in Fig. 3.

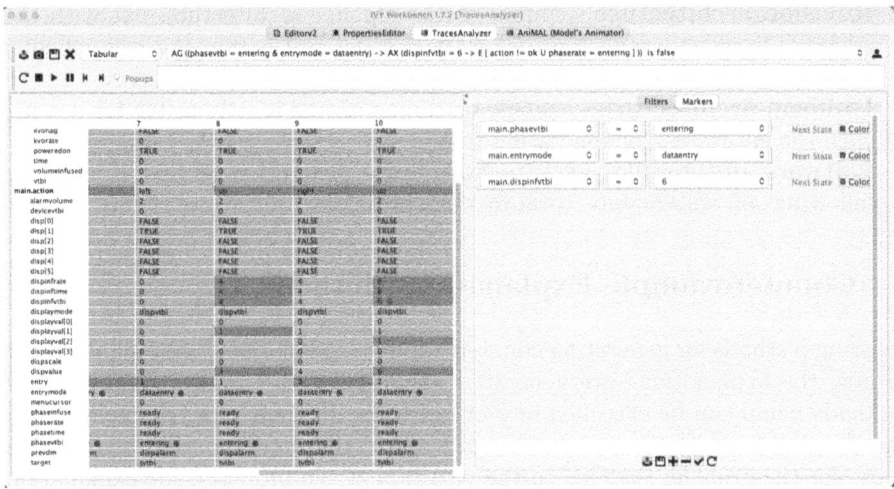

Fig. 3. IVY Workbench's ability to visually highlight parts of the tabular representation of the counterexample trace according to given conditions.

3.2 Determining Probable Causes of the Violation

These explanation generation methods focus on determining the probable causes that lead to the violation of the property being verified. Examples of these types of explanation generation methods include:

- Determining what sections of code can lead to the generation of deadlocks [28], where the analysis of a series of "good" and "bad" traces of the execution of a program is done to determine which parts of the trace are common in the "bad" traces and rare in the "good" traces, allowing for the isolation of anomalies within the code that lead to the deadlock.
- Utilizing a specific notion of causality [22] to deduce the most likely cause and context that lead to the violation of a probabilistic property [18]. This is done by determining the different contexts that can lead to the counterexample being generated for the given property, calculating the probabilities associated with each context, and then returning the highest probability context as the cause of the counterexample occurring.

3.3 Explanation of the Property Being Violated

These explanation generation methods focus on generating an explanation for the property that was violated, using the trace as an aid for this explanation.

To achieve this, these methods have an additional restriction in that they assume a known domain of analysis and, with it, a restricted number of properties with specific known characteristics that will be verified. As such, these types of explanations are more fit towards verification of rules and restrictions of a given domain.

Examples of these types of explanation generation methods include:

- Explanation of the regulations that are violated by a CAD design of a railway as it is being designed [29]. This is done by first defining a controlled natural language named *RailCNL*, similar to the language used in railway regulations. When a violation of a given regulation is detected in the CAD design, the violated regulation is transformed into a *RailCNL* phrase and attached to the objects in the layout that violate it. Then, these objects are highlighted in red on the current design, and when inspected will display the generated *RailCNL* phrase of the regulation that is being violated by those objects.
- The *AnaCon* tool, which generates a controlled natural language explanation for a internal conflict within a normative text [2]. This is done by first writing the normative texts in a controlled natural language form, which will be automatically translated using *AnaCon* into formal language. Then, this formal representation of the normative text is verified, and if there are internal conflicts between the different rules the subsequent counterexample is translated into the same controlled natural language the normative texts were first written in. This information is then grouped with the violated clause and displayed to the user.

3.4 Discussion

The approaches mentioned above point out many different approaches to explaining counterexamples, not just by simplifying or explaining the counterexample itself but also by determining probable causes that lead to the violation occurring, as well as presenting the properties being violated. Additionally, the methods by which the explanation is shown also vary, ranging from the use of natural language to the use of graphs and tables. The methods also vary in scope of applicability, from methods that work in very specific domains and conditions to more general methods that operate over a given temporal logic.

A key takeaway of this is that knowing which property was being verified is key in interpreting a counterexample. Additionally, domain knowledge helps explain the identified problem (cf. Sect. 3.3). In the following, these two ideas will be explored and built upon.

4 The Proposed Explanation Approach

This section describes the developed approach for generating explanations of counterexamples and its associated tool support.

4.1 Target Users

The objective of this tool is to generate explanations that non experts in model checking are able to understand during the process of using model checkers in the development of a system. Specifically, the work being done here is conceptualized for the first stages of this development process that uses model checking tools, long before concretization and commercial usage of said system.

During this process, the system will be verified, changed, and verified again and again as the development occurs and the model is refined. This verification will be handled by model checking experts, but the produced results must be shown and discussed with domain experts involved in the development as well as other project stakeholders, most of which will be non experts in model checking.

The objective of this work is to produce explanations for the most relevant results of model checking, the counterexamples, as a way to simplify this process of communication with domain experts, leading to less burden on the model checking experts, who will not have to waste time creating easier to understand explanations for the results of the model checking tools, and thus leading to better productivity within the project.

4.2 Explanation Format

The produced explanations will take the form of one or more sentences written in structured natural language.

This format was chosen for its universality, i.e., the fact that anyone will be able to read and understand it, allowing for a globally applicable solution. This allows for a much easier integration of the solution with a given project, as it avoids specialized explanation formats that would require time to adjust to and understand before being able to gain advantages from them.

It should be noted, however, that this format may not always be the best means of communicating model checking results, as different domains of knowledge may have distinct means of showcasing information that are more effective than structure natural language[2]. Such research on different types of explanations for different domains will be left as future work.

4.3 Explanation Generation

The explanation approach developed in this work requires four different inputs and, from them, produces a structured natural language explanation. The relevant inputs are:

- The formal model—the specific model over which the properties are verified.
- The results of the verification—pairs of properties and counterexample traces, where each property's verification failed, and the trace was the result produced by the model checker.
- A properties dictionary—pairs of property patterns and explanation templates, used to specify how different property types will be explained.
- Domain knowledge information—containing information about the domain that is relevant to understand the model and generate the explanations.

The approach aims to generate an explanation for each property/counterexample pair within the verification results input. For each property, it first finds a matching property pattern in the properties dictionary. When the matching succeeds, the explanation template attached to that property pattern is used to produce the explanation.

The property pattern defines what types of properties will be explained using the explanation template it is paired with. The explanation templates are natural language sentences enriched with *tokens* representing points where information (for example, some model attribute used in the property) should be inserted. During the explanation generation step, these templates will be filled in with data from the various data sources (model, property, domain knowledge) to generate an explanation for the property's counterexample. Domain information can be used to obtain adequate descriptions of the different model elements to be inserted into the traces' explanations.

A language has been defined to write the tokens, supporting querying the property, trace and domain knowledge for relevant information. The explanation is then produced by replacing the tokens in the selected template with information derived from the various inputs, as required.

This process repeats for each pair within the verification results input and results in an explanation being generated for each of them. An example of an explanation template and how it is used to create an explanation is provided in Sect. 5.2.

---

[2] For example, a system designer might prefer a system diagram where the error is highlighted rather then a series of sentences.

4.4 Tool Support

This approach was realised as an algorithm written as a Python script, where each input is a different file. These files are obtained from various sources and data from a project to generate the final explanation, as shown in Fig. 4.

Regarding the formal model, information is needed regarding which attributes and actions have been declared (and their types). Additionally, the MAL interactors language provides the possibility of defining named expressions to simplify the writing of axioms, and since these can be used within CTL properties, information regarding these definitions is also needed in the algorithm. The model was realised as a JSON object, where this information is located.

The results of the verification process are realised as a file containing a series of pairs of items where information about properties and counterexamples is located. Each item consists of a property and a counterexample, such that the property, having been violated in the given model[3], produced the counterexample it is paired with. In the current iteration, it is assumed that each property is written in CTL, and each counterexample was produced by the NuSMV model checker.

The dictionary that specifies how different property types will be explained is realised as the dictionary file, written as a series of pairs of items, and where information about how to generate explanations for different types of properties is located. Each item consists of a property pattern and an explanation template, as described in the previous section.

The information regarding the domain was realised as the domain file, again written as a series of pairs of items, where information about how to translate parts of the model is provided. As explained above, each item has a capturing pattern and an explanation template. The capturing pattern defines which element or expression is being described, while the explanation template will be used to explain that specific element or expression in the overall property explanation. The purpose of this is to allow the usage of domain-specific knowledge associated with the various names and properties of the model, allowing for a more detailed explanation of what is happening that takes into account the knowledge of the area related to the model and its properties.

5 An Illustrative Example

To demonstrate the approach, we will consider the model of an ice cream machine represented by the state machine diagram in Fig. 5. Further, we will consider that, within the various requirements that it must meet, one states that *the ice cream machine's tank temperature must be no higher than zero degrees Celsius*, with the purpose of ensuring that the ice cream remains fit for consumption.

---

[3] Specifically, the model whose information is provided in the model file.

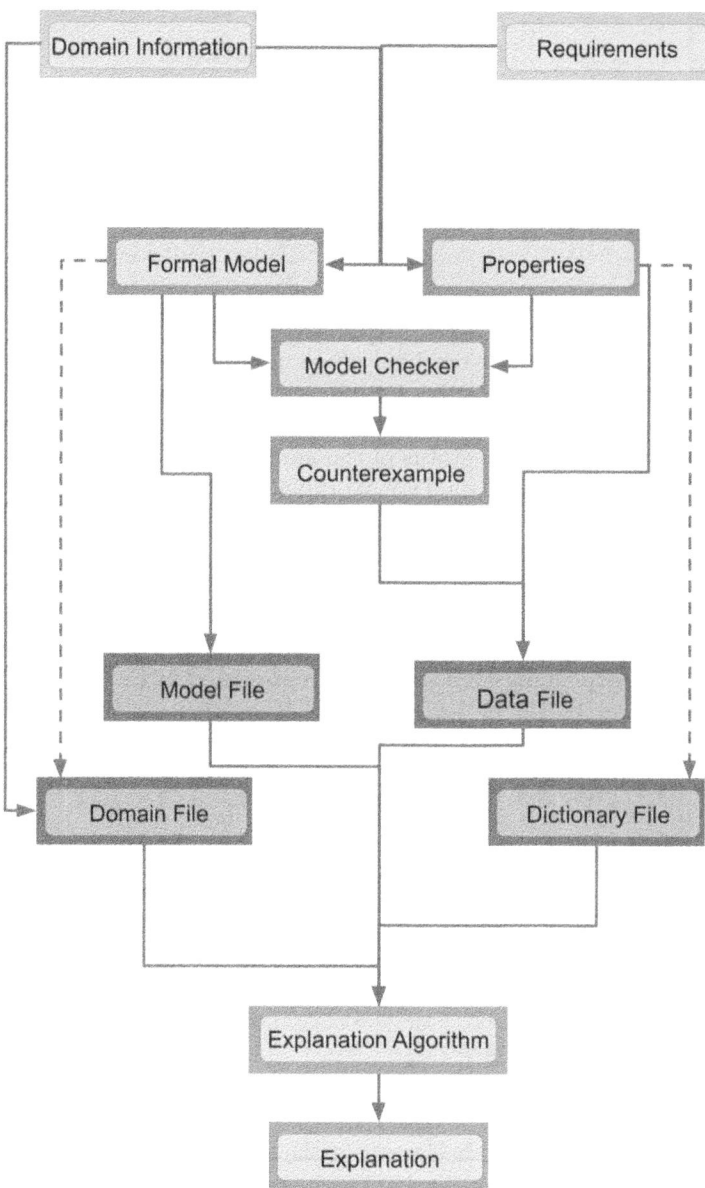

Fig. 4. The flowchart for the generation of explanations. In grey are the parts of the process that the program has no control or influence over; in blue is the realm of model checking over which the program was developed; in red are the input files for the explanation program; in green is the algorithm the program uses to generate the explanations, as well as the final explanation (Color figure online)

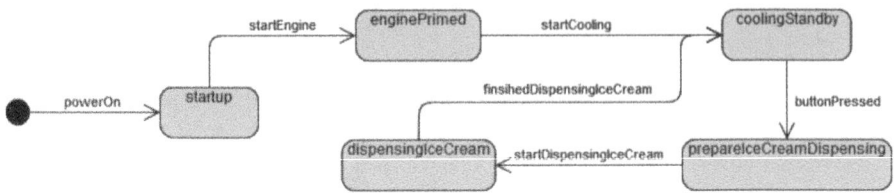

Fig. 5. A state machine diagram of the ice cream machine

5.1 Modelling and Verification

To verify that the ice cream machine obeys this requirement through model checking, the first step is to create a formal model representing the machine. For simplicity's sake, we will consider a model that only captures the temperature of the tank and the operating status of the machine.

The model was built with the IVY Workbench tool, using the MAL interactors language (see [9] for a description of the language and tool). For the present discussion, it is enough to know that the model will have two typed attributes and actions corresponding to the different transitions in Fig. 5. One attribute (tankTemperature) represents the temperature in the tank and is of type Temp (a value ranging between -10 and +10), the other (operating) is a boolean. The actions have self-explaining names. This is represented in the MAL interactors language as:

```
types
  Temp = -10..10

interactor main
  attributes
    tankTemperature: Temp
    operating: bool
  actions
    [vis] powerOn buttonPressed
    startEngine startCooling awaitButton dispensingCream
        finishedDispensing
  axioms
    ...
```

The axioms that describe the behaviour of the system are omitted for brevity.

Next, temporal logic properties that represent the machine's different requirements must be created. The CTL property AG(tankTemperature <= 0) captures the requirement that the tank temperature must be no higher than zero degrees.

Now, consider that this property was verified over the formal model of the ice cream machine and that the property was violated by said model, with the model checker producing the following counterexample trace, which illustrates a behaviour where the temperature goes above zero:

```
-> State 1.1 <-
    action = nil
    operating = false
    tankTemperature = -1
-> State 1.2 <-
    action = powerOn
    operating = false
    tankTemperature = -1
-> State 1.3 <-
    action = startEngine
    operating = true
    tankTemperature = 1
```

The goal is to explain this trace.

5.2 Explaining the Result

The developed tool must be provided with the inputs identified in Sect. 4 to generate the explanation. The model is the MAL model described above, and the verification results file will contain the CTL property and counterexample pair just described. Additionally, a domain knowledge file must created, where the meaning of the `tankTemperature` attribute will be explained. We will consider the following definition[4]:

```
tankTemperature
----
current temperature of the ice cream tank
```

Finally, the dictionary of properties is also provided.

The tool will go through the provided properties dictionary to determine how to generate the explanation. As already explained, each dictionary item is a pair consisting of a CTL property matching pattern and an explanation template. The left side of the pair provides a pattern for a CTL formula, with special tokens used to match the property being processed. If this match is successful, the right side of the dictionary item's pair is used to generate the explanation for the given data item.

For our example, consider that the dictionary contains the following pair, which will match the property given as an example previously:

```
AG({E}{1,m,lte,x,y})
----
The {DSE}{-,{E}{1,m,-,x}} was expected to always be lower than
{E}{1,m,-,y}.
However, in state {S}{1,invert,sv}, it was determined that the
{DSE}{-,{E}{1,m,-,x}} was higher than {E}{1,m,-,y}.
```

Tokens are represented inside curly braces, with each token consisting of two pairs. The first pair identifies the token type, and the second pair contains

---
[4] This is the current format for representing pairs of items in the different input files file: the left side of the pair, a dashed line, the right side of the pair.

options that change the token's interpretation by the algorithm. This applies to both the CTL pattern and the explanation template. For example, the token {E} represents (matches with) an expression. The options in {1,m,lte,x,y} then define which type of expression and how the matching will be done. In this case, the third to fifth parameters define that it is a "less than or equal" binary expression, that the left and right side of the expression will be saved in variables x and y, respectively. The first parameter attributes an identifier to the match so that the values of x and x can be later recovered. The second parameter identifies the context of the token and was included to support future developments in the approach. In the current case, x will have the value tankTemperature, and y the value 0.

The above pattern/template pair assumes a single type of explanation will be adequate whenever the property fails. For cases where a property might fail in multiple ways, it is possible to use conditions to attach multiple explanations to a single property pattern.

For example, if we extend the previous property to become AG (tankTemperature <= 0 & tankTemperature >= -5), there are now two ways it can fail: tankTemperature >= 0 is false or tankTemperature <= -5 is false. If a different explanation is required for each of these cases, then a condition attached to the dictionary property must be added for each of them:

```
AG({E}{1,m,lte,x,y} & {E}{2,m,gte,w,z})
//// {S}{1,invert,m,!({E}{1,m,-,x}) <= {E}{1,m,-,y})} ---- The
    {DSE}{-,{E}{1,m,-,x}} was greater than {E}{1,m,-,y}.
//// {S}{1,invert,m,!({E}{2,m,-,w}) >= {E}{2,m,-,z})} ---- The
    {DSE}{-,{E}{2,m,-,w}} was lower than {E}{2,m,-,z}.
;;;;
```

On the explanation template side, the token {S} refers to a state within the counterexample, and the options {1, invert, sv} tell the algorithm, respectively, that it must select the first state (1), relative to the end of the trace (invert), and that it must return the whole state value (sv). In the current case, the token with those options will be replaced with the value 1.3.

The token {E}{1,m,-,x} is used to query the first expression match on the CTL property and return the value of x (in this case, tankTemperature).

The {DSE} token is a special case, as it is used to query the domain knowledge information, replacing some piece of information in the model with its domain description. To do this replacement, the second option of this token contains the information that will be searched for within the domain file. If a match is found, the information associated with the item in the domain file will be used to replace this token. Hence {DSE}{-,{E}{1,m,-,x}} will be replaced by the domain knowledge description of the tankTemperature attribute.

With the selected explanation template, the algorithm will start replacing the various tokens to produce the final explanation. Considering all of the above, the final explanation produced by the algorithm for the given data item pair is[5]:

---

[5] Square brackets are used herein to highlight text that replaced tokens within the template.

```
The [current temperature of the ice cream tank] was expected to
always be lower than [0].
However, in state [1.3], it was determined that the [current
temperature of the ice cream tank] was higher than [0].
```

The sentence explains the reason why the property failed and points the reader to the relevant states and attributes that are causing the problem. In realistic cases, where models might have tens or more attributes and traces of tens or more states, this will be useful not only to domain experts but also to model checking experts by focusing their attention on the relevant aspects of the model and trace.

6 Conclusion and Future Work

As it currently stands, the tool, created in accordance with the presented approach, is able to produce structured natural language explanations for property/counterexample pairs that are able to locate where the violations in the property occurred in the counterexample.

6.1 Current Limitations

Despite this, there are still a number of limitations with the proposed solution.

The first is a limitation in scope, in that the proposed solution only focuses on generating explanations for the output of model checkers. While this has value, it will not fully solve the accessibility problem that these tools have, as the process of creating models and properties is still a specialized task that can only be achieved by model checking specialists.

The second is a limitation in automation, where the proposed solution requires large amounts of manual labour to achieve the desired explanations. This labor focuses on the creation of the dictionary and domain files, for whose generation there currently is no automated support, leading to a time consuming process of creating each matching pattern and attached template for both files.

6.2 Future Work

Future work will proceed along three main lines: validating the approach, further developing the approach, and exploring recent developments in large language models.

Validation of the Approach. One immediate focus is on validating the method to generate Natural Language explanations. We plan to carry out empirical studies with domain experts to evaluate the quality of the generated explanations.

These studies will likely involve answering two types of questions: the first regarding the understandability of the generated explanations (i.e., can the test subjects understand the explanations that are being generated); the second regarding the usefulness of the explanations (i.e., given a generated explanation, can test subjects determine what the problem with the formal model is).

At a later stage, if different types of explanations are developed (see below), questions to determine the understandability and usefulness of those different types of explanations will also be asked, as well as rankings between different types of explanations and evaluations of the usefulness of presenting multiple different types of explanations simultaneously.

Further Development. The evaluation results will be used to drive further development of the approach. At this moment, several possible lines of further development work have been envisaged:

- **Generation of different types of explanations.** A possibility for further development of the approach is to generalise the explanation method to allow for different types of explanations to be created. For example, an alternative to the structured natural language explanation might be using a table representation of the counterexample and colouring relevant parts within it to highlight them (cf. the IVY workbench tool).
  To do this in an automated manner, an approach to identify, for a given type of CTL property, where the relevant parts can be found in the counterexamples will have to be developed. If this is feasible, it will be possible to have multiple different types of explanations for the same property/counterexample pair.
  This may additionally allow for the creation of links between a given domain and types of explanations that would be best suited to explain the problem to experts within that domain, leading to greater understanding and effectiveness of the explanations.
- **Automatic input file generation.** At the moment, the various files the tool needs still have to be prepared manually. One obvious improvement is to automate the generation of these files.
  This generation would be done through reading the model file, and from its various parts generating at least partially each input file for the approach.
  For the data file, the properties that will be verified are in the model, and so can be added directly to the data file, while the counterexamples may be obtained from running the model checker and putting the verification results within the data file.
  Regarding the dictionary file, the goal is to have a predefined collection of generic patterns, in the style of the IVY tool. Nevertheless, in cases where a pattern is not found that matches a property, we can automatically generate a matching pattern for it and add it to the dictionary file. This is doable by converting the grammar rules of the property into the corresponding tokens, effectively reversing the matching process that occurs within the program. The attached template, however, has no means of being properly generated automatically and it will have to be edited manually when the property is first found.

For the domain file, an idea for this is to use an approach similar to literate programming [26] and use the comments in the model to assign meaning to different relevant elements. This would allow for domain data to be added automatically to the model, and allow for its use within the dictionary.
- **Using ontologies to express domain information**
  Another avenue that will be explored in regards to domain knowledge is the usage of ontologies as a means of storage and exploration of the domain, with the objective of allowing for better use of domain knowledge within the explanations.
  Ontologies are databases that denote the relations of different objects within a given domain, allowing for the exploration of these in a program. This allows for explanations to relate names and concepts expressed within the model with other objects of the same domain, allowing the creation of explanations with greater domain context.
  However, while there is work related to usage of ontologies to produce counterexample explanations [17], there is still a need to explore this topic to determine if the value gained from implementing them within the proposed solution would be worth the effort required to integrate them in the approach.
- **Integration with the IVY Workbench.** A final direction for further development involves integrating the tool with the IVY Workbench. The idea is to integrate the tool within the counterexample generation part of IVY, and use the approach to generate an explanation for each produced counterexample. Additionally, if generating multiple representations is found viable, an explanation may be added to the tabular representation within IVY to improve the process of understanding the counterexample.

Exploring Large Language Models. Given recent developments in Artificial Intelligence, in particular in large language models (LLM) [7] and code summarisation (cf. [34]), it is relevant to explore how these might influence the generation of explanations. This can range from using LLM to improve the generated explanations to exploring the generations of explanations directly from the model and traces.

References

1. Allen, J.F.: Natural Language Processing, pp. 1218–1222. Wiley, Hoboken (2003)
2. Angelov, K., Camilleri, J.J., Schneider, G.: A framework for conflict analysis of normative texts written in controlled natural language. J. Logic Algebraic Program. **82**(5), 216–240 (2013). https://doi.org/10.1016/j.jlap.2013.03.002. Formal Languages and Analysis of Contract-Oriented Software (FLACOS'11)
3. Behrmann, G., David, A., Larsen, K.G.: A tutorial on UPPAAL. In: Bernardo, M., Corradini, F. (eds.) SFM-RT 2004. LNCS, vol. 3185, pp. 200–236. Springer, Heidelberg (2004). https://doi.org/10.1007/978-3-540-30080-9_7
4. Bjørner, D., Havelund, K.: 40 years of formal methods. In: Jones, C., Pihlajasaari, P., Sun, J. (eds.) FM 2014. LNCS, vol. 8442, pp. 42–61. Springer, Cham (2014). https://doi.org/10.1007/978-3-319-06410-9_4

5. Bolton, M.L., Bass, E.J., Siminiceanu, R.I.: Using formal verification to evaluate human-automation interaction: a review. IEEE Trans. Syst. Man Cybern. Syst. **43**(3), 488–503 (2013). https://doi.org/10.1109/TSMCA.2012.2210406
6. Bozzano, M., et al.: nuXmv 2.0.0 User Manual. FBK
7. Brown, T.B., et al.: Language models are few-shot learners (2020). https://doi.org/10.48550/ARXIV.2005.14165
8. Burch, J.R., Clarke, E.M., McMillan, K.L.: Symbolic model checking: 10^{20} states and beyond. In: Proceedings of the Fifth Annual IEEE Symposium on Logic In Computer Science, pp. 428–439. IEEE Computer Society Press (1990). https://doi.org/10.1016/0890-5401(92)90017-A
9. Campos, J.C., Harrison, M.D.: Interaction engineering using the ivy tool. In: ACM Symposium on Engineering Interactive Computing Systems (EICS 2009), pp. 35–44. ACM, New York (2009). https://doi.org/10.1145/1570433.1570442
10. Campos, J., Fayollas, C., Harrison, M., Martinie, C., Masci, P., Palanque, P.: Supporting the analysis of safety critical user interfaces: an exploration of three formal tools. ACM Trans. Comput.-Hum. Interact. **27**(5) (2020). https://doi.org/10.1145/3404199
11. Campos, J., Sousa, M., Alves, M., Harrison, M.: Formal verification of a space system's user interface with the ivy workbench. IEEE Trans. Hum.-Mach. Syst. **46**(2), 303–316 (2016). https://doi.org/10.1109/THMS.2015.2421511
12. Cimatti, A., et al.: NuSMV 2: an OpenSource tool for symbolic model checking. In: Brinksma, E., Larsen, K.G. (eds.) CAV 2002. LNCS, vol. 2404, pp. 359–364. Springer, Heidelberg (2002). https://doi.org/10.1007/3-540-45657-0_29
13. Cimatti, A., et al.: Formal verification and validation of ERTMS industrial railway train spacing system. In: Madhusudan, P., Seshia, S.A. (eds.) CAV 2012. LNCS, vol. 7358, pp. 378–393. Springer, Heidelberg (2012). https://doi.org/10.1007/978-3-642-31424-7_29
14. Clarke, E.M., Emerson, E.A., Sistla, A.P.: Automatic verification of finite-state concurrent systems using temporal logic specifications. ACM Trans. Program. Lang. Syst. **8**(2), 244–263 (1986). https://doi.org/10.1145/5397.5399
15. Clarke, E., Kroening, D., Lerda, F.: A tool for checking ANSI-C programs. In: Jensen, K., Podelski, A. (eds.) TACAS 2004. LNCS, vol. 2988, pp. 168–176. Springer, Heidelberg (2004). https://doi.org/10.1007/978-3-540-24730-2_15
16. Clarke, E.M.: The birth of model checking. In: Grumberg, O., Veith, H. (eds.) 25 Years of Model Checking. LNCS, vol. 5000, pp. 1–26. Springer, Heidelberg (2008). https://doi.org/10.1007/978-3-540-69850-0_1
17. Crapo, A., Moitra, A., McMillan, C., Russell, D.: Requirements capture and analysis in assert(tm). In: 2017 IEEE 25th International Requirements Engineering Conference (RE), pp. 283–291 (2017). https://doi.org/10.1109/RE.2017.54
18. Debbi, H., Bourahla, M.: Generating diagnoses for probabilistic model checking using causality. J. Comput. Inf. Technol. **21**(1), 13–22 (2013). https://doi.org/10.2498/cit.1002115
19. Duflot, M., Kwiatkowska, M., Norman, G., Parker, D.: A formal analysis of Bluetooth device discovery. In: Proceedings of the 1st International Symposium on Leveraging Applications of Formal Methods (ISOLA 2004) (2006). https://doi.org/10.1007/s10009-006-0014-x
20. Gatt, A., Krahmer, E.: Survey of the state of the art in natural language generation: core tasks, applications and evaluation. J. Artif. Intell. Res. **61**(1), 65–170 (2018). https://doi.org/10.1613/jair.5477

21. Groce, A., Kroening, D.: Making the most of BMC counterexamples. Electron. Notes Theor. Comput. Sci. **119**(2), 67–81 (2005). https://doi.org/10.1016/j.entcs.2004.12.023. Proceedings of the 2nd International Workshop on Bounded Model Checking (BMC 2004)
22. Halpern, J.Y., Pearl, J.: Causes and explanations: a structural-model approach. Part I: Causes. Br. J. Philos. Sci. **56**(4), 843–887 (2005). https://doi.org/10.1093/bjps/axi147
23. Harrison, M.D., et al.: Formal techniques in the safety analysis of software components of a new dialysis machine. Sci. Comput. Program. (2019). https://doi.org/10.1016/j.scico.2019.02.003
24. Holloway, C.: Why engineers should consider formal methods. In: Proceedings of the 16th DASC. AIAA/IEEE Digital Avionics Systems Conference. Reflections to the Future, vol. 1, pp. 1.3–16 (1997). https://doi.org/10.1109/DASC.1997.635021
25. Holzmann, G.: The SPIN Model Checker: Primer and Reference Manual, 1st edn. Addison-Wesley Professional (2011)
26. Knuth, D.E.: Literate programming. Comput. J. **27**(2), 97–111 (1984). https://doi.org/10.1093/comjnl/27.2.97
27. Kwiatkowska, M., Norman, G., Parker, D.: PRISM 4.0: verification of probabilistic real-time systems. In: Gopalakrishnan, G., Qadeer, S. (eds.) CAV 2011. LNCS, vol. 6806, pp. 585–591. Springer, Heidelberg (2011). https://doi.org/10.1007/978-3-642-22110-1_47
28. Leue, S., Tabaei Befrouei, M.: Counterexample explanation by anomaly detection. In: Donaldson, A., Parker, D. (eds.) SPIN 2012. LNCS, vol. 7385, pp. 24–42. Springer, Heidelberg (2012). https://doi.org/10.1007/978-3-642-31759-0_5
29. Luteberget, B., Camilleri, J.J., Johansen, C., Schneider, G.: Participatory verification of railway infrastructure by representing regulations in RailCNL. In: Cimatti, A., Sirjani, M. (eds.) SEFM 2017. LNCS, vol. 10469, pp. 87–103. Springer, Cham (2017). https://doi.org/10.1007/978-3-319-66197-1_6
30. Ovsiannikova, P., Buzhinsky, I., Pakonen, A., Vyatkin, V.: Oeritte: user-friendly counterexample explanation for model checking. IEEE Access **9**, 61383–61397 (2021). https://doi.org/10.1109/ACCESS.2021.3073459
31. Perera, R., Nand, P.: Recent advances in natural language generation: a survey and classification of the empirical literature. Comput. Inform. **36**(1), 1–32 (2017). https://doi.org/10.4149/cai_2017_1_1
32. Pnueli, A.: The temporal logic of programs. In: 18th Annual Symposium on Foundations of Computer Science (SFCS 1977), pp. 46–57 (1977). https://doi.org/10.1109/SFCS.1977.32
33. Reiter, E., Dale, R.: Building applied natural language generation systems. Nat. Lang. Eng. **3**(1), 57–87 (1997). https://doi.org/10.1017/S1351324997001502
34. Tufano, R., Pascarella, L., Bavota, G.: Automating code-related tasks through transformers: the impact of pre-training. Paper Accepted at ICSE 2023 (2023). https://doi.org/10.48550/arXiv.2302.04048
35. Woodcock, J., Larsen, P.G., Bicarregui, J., Fitzgerald, J.: Formal methods: practice and experience. ACM Comput. Surv. **41**(4) (2009). https://doi.org/10.1145/1592434.1592436

Merging Creativity with Computation in Sketch-to-Code Transitions

Tommaso Calò[✉]

Politecnico di Torino, Turin, Italy
tommaso.calo@polito.it

Abstract. Artificial Intelligence (AI) has been an active research area for a long time, but its adoption as a mainstream technology in the creative/design space is a relatively new phenomenon. Much of the research has been directed into creating systems capable of performing tasks, and while many powerful systems and applications have emerged, the user experience has not kept pace. In applications of AI systems to the design process, there has been a general lack of focus on the designers creativity. Most of the current approaches focus on the use of Artificial Intelligence to replace the repetitive work in the design process while ignoring the subversive changes in thinking and methodology that AI can bring to designers' as end users. This dissertation work explores the opportunities for AI to bring radical changes in human experiences during the design and creative processes. Our most recent work focuses on the application of sketch-to-code algorithms and their adaptation to assist designers' creative ability and provide new approaches to augment design cognition.

1 Context and Motivation

Despite increasing levels of automation enabled by Artificial Intelligence - whether it is AI driving our vehicles, designing our drugs, determining what news and information we see, and even deciding how our money is invested - the common thread among these systems is the human element. AI's long-term success is contingent upon our acknowledgment that people are critical in its design, operation, and use [6,13,30].

Human-Centered AI (HCAI) [25] is an emerging discipline with the intent of creating AI systems that amplify and augment rather than displace human abilities [14]. HCAI seeks to preserve human control in a way that ensures artificial intelligence meets our needs while also operating transparently, delivering equitable outcomes, and respecting privacy [15].

Advocates of this new synthesis seek to amplify, augment, and enhance human abilities so as to empower people, build their self-efficacy, support creativity, recognize responsibility, and promote social connections. Researchers, developers, business leaders, policymakers, and others are expanding the technology-centered scope of AI to include HCAI ways of thinking. This expansion from an algorithm-focused view to embrace a human-centered perspective can shape

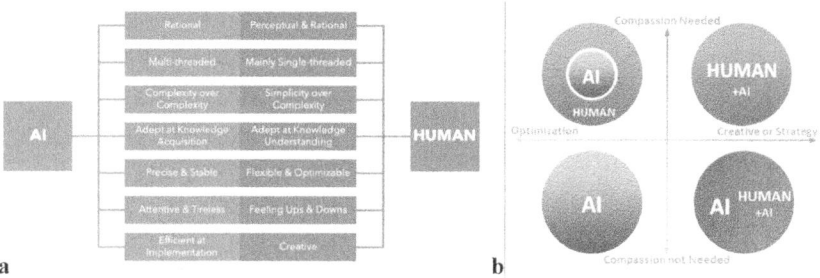

Fig. 1. (a) Modalities by which Human-AI complement each other, (b) Blueprint of Human-AI collaboration. [29]

the future of technology so as to better serve human needs. Educators, designers, software engineers, product managers, evaluators, and government agency staffers can build on AI-driven technologies to design products and services that make life better for their users [28].

AI creativity refers to the ability of humans and AI to co-live and co-create by playing to each other's strengths to achieve more. AI should complement to human intelligence, and it consolidates wisdom from human supervisions, making collaboration across time possible. AI empowers humans throughout the entire creative process and makes creativity more accessible and more inclusive than ever.

AI has been shown to be suitable to deal with repetitive and predictable problems, as well as complex and multi-tasking scenarios; while humans are more flexible and creative, and adept at knowledge understanding and strategic thinking, as is summarized in (Fig 1). Collaboration between humans and AI varies across domains [1,29] . Human leads where tasks are more about creative or strategy and compassion is needed, while AI leads where tasks are more about routine or optimization and compassion is not needed (Fig. 1).

My dissertation work aims to exploit recently discovered AI methods to enhance human creativity in the design process, analyze existing approaches' opportunities and limitations, and develop, test, and evaluate new techniques and tools to enhance designers' creativity. As a first contribution, we describe a novel approach to enhance designer creativity in automated sketch-to-code transition. This paper's initial section presents an overview of Human-Centered AI and its importance in augmenting human abilities, emphasizing the human aspect in AI's design and application. In Sect. 2 the paper delves into various studies and projects demonstrating AI's successful application in enhancing human creativity within the design domain. This background sets the context for the paper's focus on AI's role in creative processes. Section 3 outlines the specific research objectives, detailing the aim to leverage AI to enhance creativity in design and the research approach, including exploring existing applications and new tools and formulating guidelines for AI-enhanced design processes. Section 5 discusses the results and contributions made so far, showcasing progress in integrating AI into design tools, and concludes with a reflection on the current status and long-term goals, including further research into AI applications and the development of empowering guidelines for designers' creativity.

2 Background and Related Works

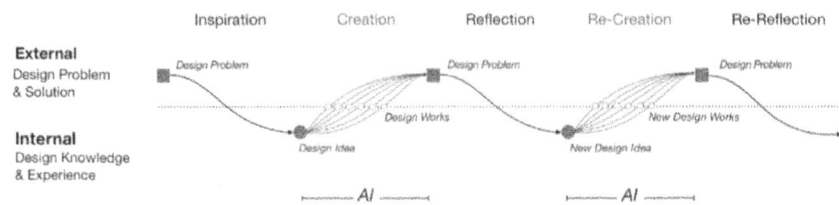

Fig. 2. The AI-augmented creative design cycle as described by [32]. The authors describe how AI can be used to enhance the design process by generating diverse creations guided by the users' input. Allowing designers to explore a wider solution space. Figure demonstrate how the cycle of continuous interaction between creation and reflection remains unchanged.

More and more scholars are studying the AI-powered, AI-enhanced, or AI-assisted human creativity [6,9,11,16,22], reporting the application of AI in the creative industry [2] and art industries [8]. Designers and design researchers also discussed and practiced data-driven design [20].

The work of Zeng et al. [32] investigated whether AI can be used to augment design creativity. They based their approach on theoretical design research [12,23] and adapted the existing design cycle to incorporate Artificial Intelligence (Fig. 2). They apply to typeface design for Chinese characters, that has stylistically a very diverse distribution. Modern Chinese character fonts do not exhibit this level of variation, so the authors attempt to use AI to augment typeface designers' creative abilities. They trained a generative network on standardized Chinese typefaces. At inference time the network was then used to generate many typefaces. The designers can explore the generated fonts to find ones that matched their desired features; input fonts that adversely affected the resulting sets were removed, and the network was retrained. This cycle was repeated until the results matched designer's needs. The study shows how to augment the design cycle without altering it. Designers are able to control every aspect of the design process while the AI augments the designer creative cognition by generating a diverse set of typeface forms, hypothetically larger than would be feasible for a design team to produce.

Sun et al. [26] presented a system that supports designers of digital icons. The authors trained a paired generative model [31] on a large database of icons. With this system, humans can guide the generation of icons by specifying its structure and a "style," the system colorizes the contours according to the chosen style.

Not to overwhelm the user with a large number of possible varieties, the system randomly proposes a selection of styles to the user. Even considering that the output is relatively simple icons, a subjective evaluation study conducted by the authors confirmed that the method performed best than a selection of other methods.

Another successful application of AI in the designer processes proposed by Zhang et al. [21] is system for generating user-guided high-quality magazine layouts. The network was trained on a large data set of magazine layout with semantic annotations with associated keyword-based summaries of textual content. Users can exercise control over the layout generation process by roughly sketching out elements on the page to indicate approximate positions and sizes of individual elements.

Another system to enhance design ideation has been proposed by Chen et al. [7]. It consists of a semantic ideation network and a visual concepts synthesis network. In the initial design session, the users can use the semantic ideation network to tune how far they would like to explore from the initial idea, for example, a user can start with the concept "spoon," and land on "straw" via "soup," and also view and filter the resulting semantic connections. In the second phase, the system uses a generative network [17] to generate novel images that attempt to synthesize a the selected visual concepts.

These and similar works demonstrate that AI can successfully enhance the designers' creative process. Hence, I want to build upon the existing knowledge, generating new strategies to improve the designers' capabilities in the creative part of their design process. My work will focus on finding under-explored cases and tasks of designers' processes in which AI could generate benefits in creating and refining existing ideas.

3 Research Objective

The overall goal of the dissertation work is to explore how to use artificial intelligence to augment design creativity, expand new design methods and design forms, and improve the quality of design. Our proposal primarily caters to designers, both novices and experts alike. For novices, the intuitive sketch-based interaction offers an easy entry point, while experts can leverage the tool to expedite the transition from design to code, allowing for a more fluid design exploration experience. Specifically, the following research questions will be addressed:

1. **RQ1:** How might we embrace AI and apply it to fostering personal human creativity in design?
2. **RQ2:** How does adjusting the AI control over the design decisions affect the human's creativity?
3. **RQ3:** What are the effects of AI over the designer's perception of limitations and frustration?
4. **RQ4:** What are the common characteristics and best practices that we can synthesize to design systems able to provide improvement in designers' creativity?

The presented research questions are synthesized from the needs highlighted in the available literature, and include hints from my personal skills and research interests, as well as long term goals of the research group.

4 Research Approach

The objective of the research is to propose and evaluate approaches that can empower designer creativity. The current research plan envisions three main phases:

1. **Exploring opportunities and limitations of existing approaches to enhance designers creativity.** An analysis of the literature is carried out with the goal of identifying the potential need for alternatives or improved integration of the existing approaches (RQ1).
2. **Development of new techniques or tools to enhance designers creativity.** We aim to design, implement, and evaluate novel HCAI approaches that enhance the creative cognition of the designers, with a specific regards to the research of systems that potentially allow an high-user control and an high automation (RQ2). Evaluation of the effect of the tools being used on the quality and quantity of the outcomes, as well as the cognitive aspect of usability (RQ3).
3. **Guidelines elicitation.** Starting from the outcomes of the previous phases, explicitly identify guidelines for the design of HCAI tools that enhance designers creativity (RQ4). Extensive evaluation of the usefulness, usability, and pleasure of use of the tools that abide by such guidelines.

The methods include identifying relevant dimensions (in terms of quality of the outcome and support for the designers), converting such dimensions into measurable variables, and conducting user testing to establish an empirical basis for the approaches being offered and for making comparisons between them.

To evaluate the proposal, we will measure the impact of the proposed solutions on design outcomes and their cognitive usability and creativity enhancement capabilities through extensive user study, which will make use of the NASA Task Load index [18] and Creativity Support Index (CSI) [10] and other ad-hoc evaluation strategies. On the AI side, common performance metrics will be used. In a subsequent phase, we will derive guidelines for the HCAI tools' design and test the usefulness, usability, and user satisfaction of tools that follow these guidelines.

5 Results and Contributions to Date

Along with my research group, we are focused on researching how Artificial Intelligence is being integrated into design support tools. One of the critical human-computer interaction considerations when creating these tools is how the end-user will communicate or interact with the system. Sketching is the

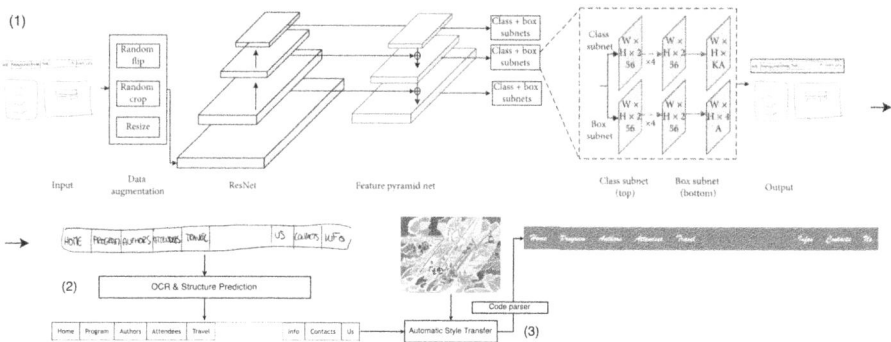

Fig. 3. (1) Starting from the sketch of a webpage, we perform a segmentation of its interface. (2) We infer the structure and the textual elements of the selected component. (3) Style properties of the reference image are extracted and injected into the structural properties of the sketch. Finally, a parser generates the final code along with the rendering of the component.

most common modality of interaction [19]. Sketch-based UI systems are familiar, comfortable, and intuitive to use for creators. Sketches can be used to provide information, commonly referred to as "hints" to the network to constrain or guide the output. When designers first start thinking about a graphical user interface (GUI), they often sketch rough pictures of the screen layouts. Designers use these sketches and other low-fidelity techniques to quickly consider design ideas, later shifting to interface construction tools or handing off the design to a programmer. However, transitioning from those sketches to a coded interface with a suitable look and feel is still a manual and time-consuming task [27] (Fig. 3).

Several research projects, indeed, have tried to automate this translation. For instance, Beltramelli [3] proposed Pix2code, an end-to-end approach based on Convolutional and Recurrent Neural Networks that allows the generation of code from a mock-up screenshot taken as input. Robinson [24], instead, presented sketch2code, a system to transform hand-drawn sketches into coded GUIs automatically. We put forward a novel approach in [4,5] where we focus on translating a sketch of a Web interface to the related code, letting the designer also choose the *style* of the generated elements from an arbitrarily chosen picture, such as an artwork or an infographic. In this way, designers can freely *explore* the space of possible solutions, maintaining complete control over the generation process. Given its complexity, the problem of style selection is split into two sub-problems: a) the selection of colors, accomplished with a clustering-based technique that extracts the most prominent colors in the reference image, and b) a feature distance-based metric technique for selecting the style of the text. We experimented the approach with a *navigation bar* because of its wide presence on many websites. Findings show that our method is able to select the style of the reference template effectively for the sketched component, showing good results in color and the font selection that well resembles the referenced style image. Within our framework, behavior and components are discerned from the segmentation of the sketched interface, laying out both its structure and textual elements. For future integration of new components, our system will be expanded to further recognize and translate additional segmented patterns within the sketches. Users interact with our system primarily through sketches, which are more than mere visual representations. These sketches serve as "hints" instructing the AI on the expected outcome. The richer the detail in the sketch, the more refined the AI's understanding of the task at hand. This provides a seamless mechanism for users to communicate their design aspirations and functional requirements. The essence of our proposal is a harmonious blend of human creativity and AI efficiency. The initial sketches, which represent the crux of the design idea, are purely human-derived. Our AI comes into play by streamlining the conversion process of these sketches into coded interfaces, while also allowing designers to infuse style elements from reference images. We recognize the value of iterative design, shaped by continual feedback. Post the AI's interpretation and translation of sketches, designers are presented with a prototype. They can then relay feedback which, when integrated, refines the AI's understanding and output, making the tool more adaptive over time. While our AI plays a pivotal role in bridging sketches to coded interfaces, the designer remains central to the entire process. From sketching the initial design, choosing reference styles, to providing feedback on the AI-generated prototype, the designer is consistently in the loop, ensuring the tool complements rather than replaces the human touch.

6 Future Work

We plan to expand the research into a more broad set of AI applications to enhance creativity, such as image and text generation. Specifically, We will focus on overcoming AI limitations to fulfill the designers needs and analyze in depth the aspect of creative processes that can be enhanced by the more advanced AI methods. The results will be used to elicit and refine guidelines to design systems that can empirically empower designers creativity. I expect our contribution to influence and bring consistent improvement to the creative process of designers and to bring novelty and knowledge progresses in the intersection between HCI and AI.

References

1. Amershi, S., et al.: Guidelines for human-AI interaction. In: Proceedings of the 2019 CHI Conference on Human Factors in Computing Systems, CHI 2019, pp. 1–13. Association for Computing Machinery, New York (2019). https://doi.org/10.1145/3290605.3300233
2. Anantrasirichai, N., Bull, D.: Artificial intelligence in the creative industries: a review. Artif. Intell. Rev. **55**, 589–656 (2022). https://doi.org/10.1007/s10462-021-10039-7

3. Beltramelli, T.: Pix2code: generating code from a graphical user interface screenshot. In: Proceedings of the ACM SIGCHI Symposium on Engineering Interactive Computing Systems, EICS 2018. Association for Computing Machinery, New York (2018). https://doi.org/10.1145/3220134.3220135
4. Calò, T., De Russis, L.: Creating dynamic prototypes from web page sketches. In: Proceedings of the 1st ACM SIGPLAN International Workshop on Programming Abstractions and Interactive Notations, Tools, and Environments. Association for Computing Machinery, New York (2022). https://doi.org/10.1145/3563836.3568724
5. Calò, T., De Russis, L.: Style-aware sketch-to-code conversion for the web. In: Companion of the 2022 ACM SIGCHI Symposium on Engineering Interactive Computing Systems, EICS 2022 Companion, pp. 44–47. Association for Computing Machinery, New York (2022). https://doi.org/10.1145/3531706.3536462
6. Carter, S., Nielsen, M.: Using artificial intelligence to augment human intelligence. Distill (2017). https://doi.org/10.23915/distill.00009. https://distill.pub/2017/aia
7. Chen, L., et al.: An artificial intelligence based data-driven approach for design ideation. J. Visual Commun. Image Represent. **61**, 10–22 (2019). https://doi.org/10.1016/j.jvcir.2019.02.009. https://www.sciencedirect.com/science/article/pii/S1047320319300604
8. Chen, W., Shidujaman, M., Xuelin, T.: Aiart: towards artificial intelligence art (2020)
9. Chen, X., Duan, Y., Houthooft, R., Schulman, J., Sutskever, I., Abbeel, P.: Infogan: Interpretable representation learning by information maximizing generative adversarial nets. In: Lee, D., Sugiyama, M., Luxburg, U., Guyon, I., Garnett, R. (eds.) Advances in Neural Information Processing Systems, vol. 29. Curran Associates, Inc. (2016). https://proceedings.neurips.cc/paper/2016/file/7c9d0b1f96aebd7b5eca8c3edaa19ebb-Paper.pdf
10. Cherry, E., Latulipe, C.: Quantifying the creativity support of digital tools through the creativity support index. ACM Trans. Comput.-Human Interact. (TOCHI) **21**, 1–25 (2014)
11. Colton, S., Wiggins, G.: Computational creativity: the final frontier? Front. Artif. Intell. Appl. **242**, 21–26 (2012). https://doi.org/10.3233/978-1-61499-098-7-21
12. Dorst, K., Cross, N.: Creativity in the design process: co-evolution of problem-solution. Des. Stud. **22**, 425–437 (2001). https://doi.org/10.1016/S0142-694X(01)00009-6
13. Dove, G., Halskov, K., Forlizzi, J., Zimmerman, J.: UX design innovation: challenges for working with machine learning as a design material (2017). https://doi.org/10.1145/3025453.3025739
14. Engelbart, D.: Augmenting human intellect: a conceptual framework (1962). https://www.bibsonomy.org/bibtex/298050040b80f71891383fcd83d7c7100/fcerutti
15. Fiebrink, R.A.: Real-time human interaction with supervised learning algorithms for music composition and performance (2011). aAI3445567
16. Fischer, G., Nakakoji, K.: Amplifying Designers' Creativity with Domain-Oriented Design Environments, pp. 343–364. Springer, Dordrecht (1994). https://doi.org/10.1007/978-94-017-0793-0_25
17. Goodfellow, I., et al.: Generative adversarial nets. In: Ghahramani, Z., Welling, M., Cortes, C., Lawrence, N., Weinberger, K. (eds.) Advances in Neural Information Processing Systems, vol. 27. Curran Associates, Inc. (2014). https://proceedings.neurips.cc/paper/2014/file/5ca3e9b122f61f8f06494c97b1afccf3-Paper.pdf

18. Hart, S.G.: Nasa-task load index (nasa-tlx); 20 years later. Hum. Fact. J. Hum. Fact. Ergon. Soc. (2006)
19. Hughes, R.T., Zhu, L., Bednarz, T.: Generative adversarial networks-enabled human-artificial intelligence collaborative applications for creative and design industries: a systematic review of current approaches and trends. Front. Artif. Intell. **4** (2021)
20. King, R., Churchill, E.F., Tan, C.: Designing with data: improving the user experience with a/b testing (2017)
21. Li, J., Yang, J., Hertzmann, A., Zhang, J., Xu, T.: Layoutgan: generating graphic layouts with wireframe discriminators (2019)
22. Miller, A.: The Artist in the Machine: The World of AI-Powered Creativity (2019). https://doi.org/10.7551/mitpress/11585.001.0001
23. Preece, J., Rogers, Y., Sharp, H.: Interaction Design: Beyond Human-Computer Interaction, 4th edn. Wiley, Hoboken (2015)
24. Robinson, A.: Sketch2code: generating a website from a paper mockup (2019)
25. Shneiderman, B.: Bridging the gap between ethics and practice: guidelines for reliable, safe, and trustworthy human-centered AI systems. ACM Trans. Interact. Intell. Syst. **10**(4) (2020). https://doi.org/10.1145/3419764
26. Sun, T.H., Lai, C.H., Wong, S.K., Wang, Y.S.: Adversarial colorization of icons based on contour and color conditions. In: Proceedings of the 27th ACM International Conference on Multimedia, MM 2019, pp. 683–691. Association for Computing Machinery, New York (2019). https://doi.org/10.1145/3343031.3351041
27. Walker, M., Takayama, L., Landay, J.A.: High-fidelity or low-fidelity, paper or computer? choosing attributes when testing web prototypes. Proc. Hum. Fact. Ergon. Soc. Ann. Meet. **46**(5), 661–665 (2002). https://doi.org/10.1177/154193120204600513
28. Ware, M., Frank, E., Holmes, G., Hall, M., Witten, I.H.: Interactive machine learning: letting users build classifiers. Int. J. Hum Comput Stud. **55**(3), 281–292 (2001). https://doi.org/10.1006/ijhc.2001.0499
29. Wu, Z., Ji, D., Yu, K., Zeng, X., Wu, D., Shidujaman, M.: AI creativity and the human-AI co-creation model, pp. 171–190 (2021). https://doi.org/10.1007/978-3-030-78462-1_13
30. Yang, Q.: Machine learning as a UX design material: how can we imagine beyond automation, recommenders, and reminders? (2018)
31. Yi, Z., Zhang, H., Tan, P., Gong, M.: Dualgan: unsupervised dual learning for image-to-image translation. In: 2017 IEEE International Conference on Computer Vision (ICCV), pp. 2868–2876 (2017)
32. Zeng, Z., Sun, X., Liao, X.: Artificial intelligence augments design creativity: a typeface family design experiment. In: Marcus, A., Wang, W. (eds.) HCII 2019. LNCS, vol. 11584, pp. 400–411. Springer, Cham (2019). https://doi.org/10.1007/978-3-030-23541-3_29

UX Data Visualization: Supporting Software Professionals in Exploring Users' Interaction Data

Maylon Macedo[(✉)] and Luciana Zaina

Federal University of São Carlos, Sorocaba, São Paulo, Brazil
macedomaylon@gmail.com, lzaina@ufscar.br

Abstract. The definition and understanding of UX data and its possibilities for analysis and production of insights play a key role in software teams' adoption of UX practices. Information Visualization (InfoVis) offers approaches to assist the design of visualizations that users could better explore. Besides, it provides principles to support the analysis of information that emerged from data. This paper introduces a doctoral research project that uses Design Science Research (DSR) as a general methodology to guide successive research cycles involving investigations in the grey literature and exploratory studies to build a framework. This doctoral research aims to develop a framework for assisting software teams in creating, configuring, and using visual representations to organize, extract, and communicate information embedded within UX data.

Keywords: User eXperience · UX data · HCI · Visualization · InfoVis · Design Science Research

1 Introduction and Motivation

User Experience (UX) is an important aspect of software quality that affects the product's acceptance by users [19,26]. All aspects related to the interaction between users and software fall under the field of UX [14,25]. Norman defines UX as the user's feelings about all aspects related to their interaction with the product [25]. Data about user experience can be collected before, during and after the use of a software product. This data can support software teams to make a more user-centered decision about the software evolution and to have insights for new software products [10,14].

The literature has characterized the data produced by user interaction with the product as "UX data" [2]. UX data can be quantitative or qualitative [2], and it can be produced from the application of formal collection (e.g., questionnaires) or informal (e.g., chat with users) [10,15,20,28]. However, data collection, as well as the use of UX data, is considered a difficult activity for some software teams, particularly in contexts of resource-limited companies where there is no specific professional to handle UX (e.g., small companies and startups) [15,33].

The use of UX Data visualizations, i.e. graphical representation, can help software teams to explore data abstractions. Consequently, the visualizations provide ways to reduce the effort required to infer patterns, identify trends, outliers, and other insights of emerged data [5,6,8,31]. Information Visualization (InfoVis) provides techniques for building visualizations that can help obtain information based on the visualization of emerging data properties [7,24].

Although the literature has recognized the importance of using UX data during software development, it also presents gaps related to the construction, configuration, and use of visualizations that allow the extraction of meaningful information about UX Data. This doctoral research aims to provide a framework to support software teams in constructing, configuring, and utilizing visual representations to assist in organizing, extracting, and communicating information embedded within UX Data. This solution aims to offer support and autonomy to software teams and to encourage research into HCI techniques within the limitations and needs of each team.

This paper is organized as follow: Sect. 2 presents the related works, Sect. 3 presents the problem and research questions; Sect. 4 presents details about each step of the methodology; Sect. 5 presents details and results of and grey literature review; Sect. 6 presents details about the conduct and results of the exploratory study; Sect. 7 presents the proposal of this doctoral research and Sect. 8 presents the final considerations.

2 Related Work

The literature have explored the use of visualizations for UX data. The most common focus is on using visualizations to present data collected in usability tests with users [4,9,23,31]. Other objectives are also explored in the literature, such as visualizing user navigation on websites [5,29], presenting user evaluations of software [3] and mobile applications [36], and reconstructing user journeys with a particular product [18].

Silva et al. [31] developed the UXmood tool, an interactive dashboard with visualizations of users' emotional states. The visualizations were generated using data collected from sensors, including video, audio, interaction logs, sentiment, and eye trackers. The authors evaluated the UXmood tool with 122 participants to assess the efficiency of using visualizations to communicate the data. The results showed that the visualizations were a user-friendly tool that improved the communication of UX information.

Considering website navigation data, Buono et al. [5] developed four visualizations to assist software teams with limited UX experience in analyzing usability data. An evaluation was conducted with fifteen participants to assess team satisfaction and whether the visualizations could help identify usability issues. The results revealed that the provided visualizations assisted participants in quickly and objectively analyzing the data.

Zaidi et al. [9] proposed InteracDiff, which is a prototype visualization tool that aids in interpreting UX data collected in user tests. The visualization displays participants' responses in a predefined cycle-shaped grid and groups responses into categories related to UX quality (product attractiveness, user sentiments). Twelve participants performed tasks while interacting with the visualization tool during the evaluation. The results indicated that the tool provided a straightforward approach to visualizing UX data but required a high cognitive load from participants.

To support the analysis of usability evaluation results based on user tasks, Bernhaupt et al. [4] created three visualizations. These visualizations were constructed using Shneiderman's mantra, "Overview first, zoom and filter, then details-on-demand", and featured two levels of granularity: overview and details. A case study with ten participants demonstrated the feasibility of using visualizations to extract insights from UX data.

Karapanos et al. [18] designed the iScale tool to assist UX researchers in retrospective user studies. The tool presented two versions of the timeline chart, differing in how chronological data was loaded into the visualization (automatically by the tool or through user interaction). The effectiveness of the two versions of the tool in obtaining user experience reports was evaluated by 48 participants. The results showed that using visualizations to organize collected data helped participants analyze the perceived user experience.

3 Research Goals

Related work presents a set of solutions for visualizing UX data; however, most of them are focused on building tools for specific contexts. Most of the literature does not provide ways to make flexible the application of the proposal in other contexts. [4,5,9,23,29,31]. This doctoral research differs from the related works because it does not explicitly aim to construct a visualization (or a set of them) but rather to support software teams (based on a visualization approach) in tasks to gain insights about UX from information available in their context.

We define the following search problem: *To empower software development team members with guidelines that enable them to construct and effectively use visual representations of user experience (UX) data throughout the product lifecycle.* Based on the research problem, this doctoral research aims to address the following research questions:

1. How can software teams recognize and classify different types of user interaction data?
2. Which visualization formats can assist software teams in exploring UX data to get insights to improve software or create new products?
3. How can the principles of InfoVis be employed to configure and/or design visual representations of UX data?

The goals are based on two arguments. The first is related to the gap presented by the literature regarding the lack of tools, artifacts, and methods to help extract information from UX data [10,28,33]. The second argument concerns the non-application of good practices related to the InfoVis area in visualizations to interact with UX data found in the literature [9,17,18,21].

4 Methodology

This doctoral research is guided by the Design Science Research (DSR) approach which focuses on the development and evaluation of artifacts constructed to address a specific purpose [34]. This doctoral project intends to build and evaluate guidelines and tools that will compose the framework.

DSR Cycle by Hevner [13]: The cycle defined by Hevner are *Relevance*, *Rigor*, and *Design*. In the Relevance cycle, the researcher conducts studies to include the application domain within the project scope, ensuring that the developed artifact is relevant to real needs. The Rigor cycle is responsible for exploring resources to provide knowledge about the project's scope, ensuring that the developed artifact is innovative compared to the state of the art. Finally, in the Design cycle, the construction and evaluation of artifacts take place, considering all the information captured by the Relevance and Rigor cycles.

DSR Cycles in the Doctoral Research: The organization using the DSR cycles aims to allow cross-investigation between emerging literature-based information and the discovery of needs in activities carried out in practice regarding the utilization of visual representations about UX data. In this doctoral research, the *Relevance and Rigor* cycles will always be conducted in conjunction to provide evidence and guide analytical processes based on data collected from grey literature (relevance), scientific literature (rigor), and investigations with the target audience (relevance).

A brief discussion of each step will be presented in the following paragraphs. Steps E1, E2, and IN1 are complete and present their contributions to the project, and steps E3 and E4 are in progress and thus present the expected contributions.

Review of the Scientific Literature (E1): The *objective* of this step was to investigate related works to explore the application of InfoVis techniques and uncover different types of data and visualization approaches that can provide valuable insights from UX data. The process followed the research protocol of a Systematic Literature Review (SLR) [27]. **Contribution**: Formal definitions about UX data and information about the domain of the application and the gaps in the scientific knowledge base, guaranteeing the premises of the first iteration of **relevance and rigor** cycle.

Grey Literature Review (E2): The objective of the Grey Literature Review (GLR) was to investigate the use of UX data within the software development cycle. **Contribution**: The "state of practice" that complements the scientific literature, information about which UX data are considered relevant, and which practices and tools are used to analyze them. This step represents the second iteration of the **relevance and rigor** cycle.

1st Exploratory Study (IN1): We designed five visualizations focusing on analyzing the data about a mobile application for airline ticket sales. The goal was to assess the use of these visualizations to gain insights into the issues faced by the application. **Contribution:** Data from this exploratory study is still in the process of analysis. It is expected that the qualitative analysis of the results from this study will validate the visualizations and the data used, as well as expand the definition of the necessary conditions for using these visualizations.

Meta-Analysis of Collected Data (E3): To explore and expand the extractions from literature and grey literature, a meta-analysis of the data collected in step IN1 is planned. **Expected Contribution:** To select the most suitable definitions of UX data. To explore the scope of practices used to explore UX data according to the collected reality from evaluation participants. To initiate classification regarding which visualization models are appropriate based on the nature of UX data. To provide guidance for selecting extractions that will comprise the 1st version of the framework. This step represents the third iteration of the **rigor and relevance** cycle.

Proposal of the Framework (E4): The first version of the framework aimed at exploring UX data to enhance the capabilities of professionals involved in interactive software development will be proposed. **Expected Contribution:** A framework to assist software teams in identifying and visualizing UX data. The proposal of the framework represents the third iteration of the **design** cycle of this doctoral research.

Framework Evaluation and Refinement: We planned to execute two evaluations of the framework with at least two software teams in each one. The goal is for the framework to be used by software teams for at least two development cycles so that the researcher can monitor daily the results and emerging practices of the proposal's use. **Expected Contribution:** New insights into team contexts, conditions for using the proposal, and the contributions it can make to the software development process by exploring the use of UX data. Based on the information gathered in the evaluations, the framework will be refined to incorporate new insights gained from the practical use of the proposal. This step is considered the fourth iteration of the **design** cycle, as it will produce and evaluate a refined version of the framework.

The partial results of the doctoral project will be presented in next sections.

5 Grey Literature Review

Grey literature presents information produced by practitioners and it is published on personal websites, social networks, blogs, meeting minutes, discussion forums, and question-and-answer websites [1]. Grey literature has been considered a rich source of knowledge acquired daily by software professionals. Different contributions from grey literature have been published in scientific software engineering venues, for instance, the proposal of guidelines to work with accessibility with design patterns [37], and the use of hypotheses engineering as an approach in software startups [22].

Garousi et. al. [12] proposed guidelines to investigate the grey literature systematically; in this doctoral research, we follow these guidelines to conduct a Grey Literature Review (GLR). In this doctoral research, the investigation of grey literature aimed to discover practices and challenges often adopted by practitioners during the use of UX data within the development workflow. Grey literature was included as a source of information to support the finding of "who, when, and how" UX data are used in software development practices. In next sections, the conduction and results are presented.

5.1 Conduction

During the evaluation process of grey articles that contained mentions about activities that developers perform during the construction of interactive software using UX data, as well as the difficulties they encounter in using this data was searched. Two primary research questions were formulated to differentiate the analysis between the perspective of the software team using UX data (RQ1) and the perspective of available tools and approaches (RQ2) (see Table 1).

Table 1. Research questions

RQ	Description
RQ1	What UX data have software teams explored during software development?
RQ1.1	Which professionals use UX data?
RQ1.2	What role do these UX data play in the development workflow?
RQ2	What are the approaches that have been assisting developers in the task of exploring UX data?
RQ2.1	What types of visualizations are applied to UX data?
RQ2.2	What are UX data format, and how are they used in existing approaches?

Eighteen options of two-term configuration were designed as search strings (i.e., see Table 2). The two-term configuration avoids the use of logical operators that are not common in Internet search engines and provides a way to get broader results. Google search engine was adopted due to its popularity in studies involving grey literature and being a general internet search engine [12]. No specific criteria were used to end the search; all results were subjected to inclusion and exclusion criteria. An inclusion criterion was defined as follows: (CI1) The grey article must describe the use of UX data. Three exclusion criteria were defined as follows: (CE1) duplicate content, (CE2) solely advertising content, and (CE3) grey articles in non-textual format (e.g., videos or podcasts). The links of grey articles were retrieved using the SEOQuake plugin[1].

---
[1] https://www.seoquake.com/.

The search process results in **2262 grey articles**. All the grey articles were downloaded in .PDF format to avoid losing access to the content during the research. A spreadsheet was used to organize the grey article information as follows: (i) the language of the text, (ii) a summary of the grey article content, (iii) the inclusion and exclusion criteria checking, and (iv) the link to the .PDF version of the grey article. After the inclusion and exclusion criteria application, 144 grey articles were selected. The list of the selected grey articles is available in a spreadsheet in Google Drive. After, each grey article collected was read and the extractions were associated with each RQ and stored in a form. It is not mandatory that each grey article provides contributions to all the RQs.

Table 2. Search strings used for grey literature review.

ID	Search string	ID	Search string
1	"quantitative UX"	10	"UX data" + "analysis"
2	"UX" + "data"	11	"UX data" + "charts"
3	"UX" + "Data Analysis"	12	"UX data" + "dev"
4	"UX" + "data scientist"	13	"UX data" + "information visualization"
5	"UX" + "data-driven"	14	"UX data" + "InfoVis"
6	"UX" + "information visualization"	15	"UX data" + "patterns"
7	"UX" + "data extraction"	16	"UX data" + "platform"
8	"UX" + "InfoVis"	17	"UX data" + "product development"
9	"UX data" + "evaluation"	18	"UX data" + "software development"

5.2 Results

The results are presented by answering the RQs (see Table 1). Due to the limitation of pages, we used the grey article ID listed in the spreadsheet available in the link[2] to summarize our results.

RQ1: Data-driven development is described as a continuous and iterative approach, allowing teams to validate their efforts and adjust their strategies as needed [AC71]. There is evidence that software teams have explored UX data throughout the software development process from inception to completion [AC54, AC118, AC71]. UX data can be involved in various stages, such as goal definition [AC24], hypothesis formulation [AC23], conducting tests and experiments [AC140], and analyzing results to validate or refute hypotheses [AC5, AC23, AC81]. The data-driven approach enables teams to understand user behavior better, make more informed decisions, and create experiences that are more aligned with the needs of the target audience [AC2 and AC142].

---

[2] spreadsheet with approved grey articles.

RQ1.1: *Designers* conduct user experience data analyses to understand how users interact with products, identify areas for improvement, optimize usability, and create more appealing user experiences [AC1, AC130, AC131, AC68, AC123, AC134, AC29, AC59, AC6, AC109]. *UX researchers* use data to deepen their understanding of user behavior and preferences to inform design decisions, assist in identifying usability issues, and provide insights for creating user-centered products [AC30, AC114, AC57, AC115, AC61, AC86]. *Data analysts* help transform raw data into meaningful insights, visualize relevant metrics, and provide data that supports design decisions [AC92, AC121, AC91, AC138, AC75, AC27]. *Product teams* use UX data to make informed decisions and direct efforts to meet customer needs [AC3, AC137, AC111, AC35, AC124, AC63]. *Development professionals* benefit from UX data to ensure design decisions are well-founded and products meet the expectations of the target audience [AC4, AC89, AC48, AC81, AC87, AC76]. Lastly, *product managers* use UX data to understand product performance in the market, identify growth opportunities, prioritize necessary improvements based on success metrics, and improve goal communication among team members [AC2, AC98, AC117, AC91, AC103, AC22, AC63].

RQ1.2: The importance of continuously measuring, learning, and optimizing the product based on collected data is emphasized [AC41, AC84]. UX data helps connect the development team with the target audience, enabling a more empathetic and user-centered approach [AC23, AC81]. Some essential functions of UX data can be highlighted in the development workflow: Performance, maturity, and success assessment of business goals [AC40, AC105, AC132]; Understanding user behavior [AC26, AC9, AC131, AC18, AC111, AC76]; Guiding product and experience optimization [AC1, AC130, AC139, AC132, AC85, AC100]; Identifying problems and opportunities [AC4, AC92, AC102, AC42, AC100]; and supporting decision-making [AC85, AC96].

RQ2: There are three main groups of approaches found in the grey literature, namely: **Approaches for visualizing UX data** [AC63, AC138, AC1, AC16, AC30, AC33, AC34, AC110, AC117, AC123, AC130, AC131, AC141]; **Approaches for collecting UX data** [AC55, AC33, AC40, AC84, AC117, AC41, AC110]; and **Approaches for conducting tests and research** [AC122, AC116]. A common point among the found approaches is the use of data analysis tools and the application of visualization techniques to provide information with a focus on user experience [AC27, AC49, AC101, AC143]. Approaches derived from Web Analytics help identify problematic points, such as where users are leaving the site (bounce rate), which pages have higher engagement, and which actions lead to higher conversion rates [AC71, AC118, AC58]. Table 3 shows the tools mentioned in the grey article and their respective IDs.

Table 3. Tools found on grey literature.

Tool	grey article	Tool	grey article
Google Analytics	[AC5, AC30, AC41, AC53, AC55, AC71, AC84, AC117]	W3Counter	[AC55]
		UserFeel	[AC116]
Hotjar	[AC41, AC53, AC65, AC71, AC87]	TryMyUI	[AC116]
Crazy Egg	[AC40, AC71, AC117, AC125]	Qualaroo	[AC116]
Mixpanel	[AC55, AC58, AC71, AC84]	Survey Monkey	[AC116]
Inspectlet	[AC65, AC84]	Data Studio	[AC116]
Mouseflow	[AC65]	Dovetail	[AC24]
FullStory	[AC71, AC84]	Aurelius	[AC124]
Heap	[AC71, AC84]	Ahrefs	[AC124]
UXCam	[AC58, AC65]	Lucky Orange	[AC125]
Flink	[AC71]	Optimizely	[AC87]
Lookback	[AC65]	VWO	[AC87]
Amplitude	[AC84]	lookback	[AC87]
Contentsquare	[AC65]	Userlytics	[AC87]
AppSee	[AC65, AC71, AC84]	Matomo	[AC55]

RQ2.1: The use of visualizations allows tracking performance metrics, monitoring user navigation in software, identifying behavioral patterns, and analyzing usability test results, among other applications [AC138, AC101, AC82, AC125, AC37]. There are types of visualizations that are widely accepted for analyzing UX data, such as bubble charts [AC42, AC123], scatter plots [AC34], bar charts [AC123], and tree diagrams [AC141]. Personas are recommended to visualize user needs, motivations, and behaviors [AC5, AC70, AC91, AC103, AC143]. Some types of visualization are also used as techniques for organizing and structuring data, for example, affinity diagrams, personas, empathy maps, user journey maps, and user flow maps [AC71, AC98, AC131]. Customized visualizations are also developed to represent non-standardized data, such as data generated by traffic analysis, session recordings, usability tests, surveys, conversion funnels, and user feedback, among others [AC35, AC37, AC38, AC41, AC55, AC63, AC101, AC116, AC122, AC124, AC125, AC126, AC138].

RQ2.2: UX data encompasses data collected to understand the user experience, including **qualitative**, **quantitative**, **behavioral**, and **attitudinal** data [AC40, AC96, AC16]. Qualitative data are related to the qualities and feelings of users towards the product and can include user interviews [AC57, AC89], usability tests [AC48, AC50], and observations that help understand motivations and expectations [AC85]. Quantitative data are related to numerical and measurable values, providing a more objective view of user behavior, such as conversion rate, session time, and pages visited [AC99]. Behavioral data refers to information about the actual behavior of users, such as actions, interactions, task time, and error rate [AC82, AC32]. On the other hand, attitudinal data are obtained through surveys, persona analysis, questionnaires, and interviews and are related to users' opinions, feelings, and perceptions about the software [AC79, AC63].

For a comprehensive approach it is expected to use a combination of behavioral and attitudinal data where different methods are used to obtain complementary insights and mitigate the limitations of each approach [AC66, AC86]. Integrating UX data into other approaches involves analyzing data of diverse natures; quantitative data can be analyzed using visualization tools [AC120, AC122]. In contrast, qualitative data require a more in-depth interaction to identify patterns [AC120, AC124]. Some tools and methodologies assist in integrating UX data with other approaches, such as the HEART methodology [AC136], thematic analysis [AC78], and CAQDAS tool [AC19].

6 Exploring UX Data in User Flow Data

This exploratory study aimed to gather participants' perceptions based on visualizations with regard to the usage data of a mobile application. To provide context to the participants, we presented the scenario of a mobile application, including app interfaces, usage flow, business rules, company goals, and reports of dissatisfaction from managers regarding the overall app performance.

The "reports of dissatisfaction" were elaborated considering the most common issues cited in grey literature that utilized visual representations to address the solution. The reports of dissatisfaction developed were as follows: *RD1 - Many people access TecFlyFind, but few complete the purchase*; *RD2 - The reason for the decrease in sales after February 2022 is unknown*; *RD3 - Users who like discounts cannot find TecFlyFind in internet searches*; and *RD4 - People say they do not understand how the cashback policy works*. Figure 1 shows the visualizations and the relation with the "reports of dissatisfaction".

Two visualizations were elaborated to give insights into RD1. By using a userflow chart format, the first (see Fig. 1-A) provides an overview from the user's perspective before leaving the application and thus suggests possible interfaces with serious issues. To give a supplementary perspective for RD1, the second visualization (see Fig. 1-B) in severity matrix format relates the time users spent inside the application (values within the cells) and the stage of completing the purchase at the time of exit (X-axis).

To provide insights for RD2, we use the calendar format (see Fig. 1-E) to relate access and sales data per day to events that occurred with the app's infrastructure; this visualization supports the analysis of whether the inclusion of a cashback rule alert page is the reason for the decrease in sales. A word cloud visualization (see Fig. 1-C) was developed to observe the representativeness of the terms chosen to promote the application concerning the terms searched on the internet, and this brings insights for RD3. Based on the heatmap format (see Fig. 1-D), we developed a visualization to show the areas where users click most, and this provides insights for RD4 by observing that users do not access the application's terms of use.

The data was collected from 31 participants in an online questionnaire. The data analysis is in progress. The partial results revealed that the data and visual representations can be useful for designers, stakeholders, and developers according to the participants. The results showed that the participants did not recognize that UX data can be gathered from different sources. In their perspective, UX data are generated only from traditional instruments (e.g., questionnaires) and with the intervention of UX professionals. Besides, the results restate what we saw in the literature that the definition of UX data is not clear for developers and designers

7 Toward a Framework to Design UX Information Visualization

Taking into account the partial results from the GLR and the primary study, the framework conception was started. In the context of this doctoral project, a framework is understood as a set of premises, values, and practices organized according to a structure that enables the comprehension and communication of a theme for a specific purpose [30].

The first draft of the framework followed the 5W1H method and considered the partial results obtained from GLR (see Sect. 5). The 5W1H allows the communication of information emerging from collection processes, such as the definition of taxonomies [16], ontologies [35], communication of stakeholders' viewpoints [32], and requirement extraction [11]. The method is composed of a set of elements (i.e., what, when, where, who, why, and how) that supports the explanation of elements involved with some related artifact.

Figure 2 illustrates the first draft of the framework to guide the selection of data and visualizations. Three dimensions and their respective guided questions sustain the framework: UX data (help identify and choose relevant data); developers tasks (help with how to align analyzes with the development cycle), and team (help with how to manage and handle activities involving UX data). Each dimension has a 5W1H-based structure. The 5W1H elements performed the following role:

- *What:* address the different UX types of data including quantitative data, qualitative data, behavioral data, and attitudinal data.
- *Where:* determine the data source from which UX data can be collected, such as on websites, mobile apps, field surveys, among others.
- *When:* specify when UX data collection occurs and when they are used during the user experience design process.
- *Who:* identify the professionals that can access, analyze and use the UX data.
- *Why:* point out the purpose and importance of collecting and analyzing UX data, including categories such as problem and opportunity identification, data themes and clustering analysis, product performance, user experience and navigation flow, understanding users, and stakeholder communication.

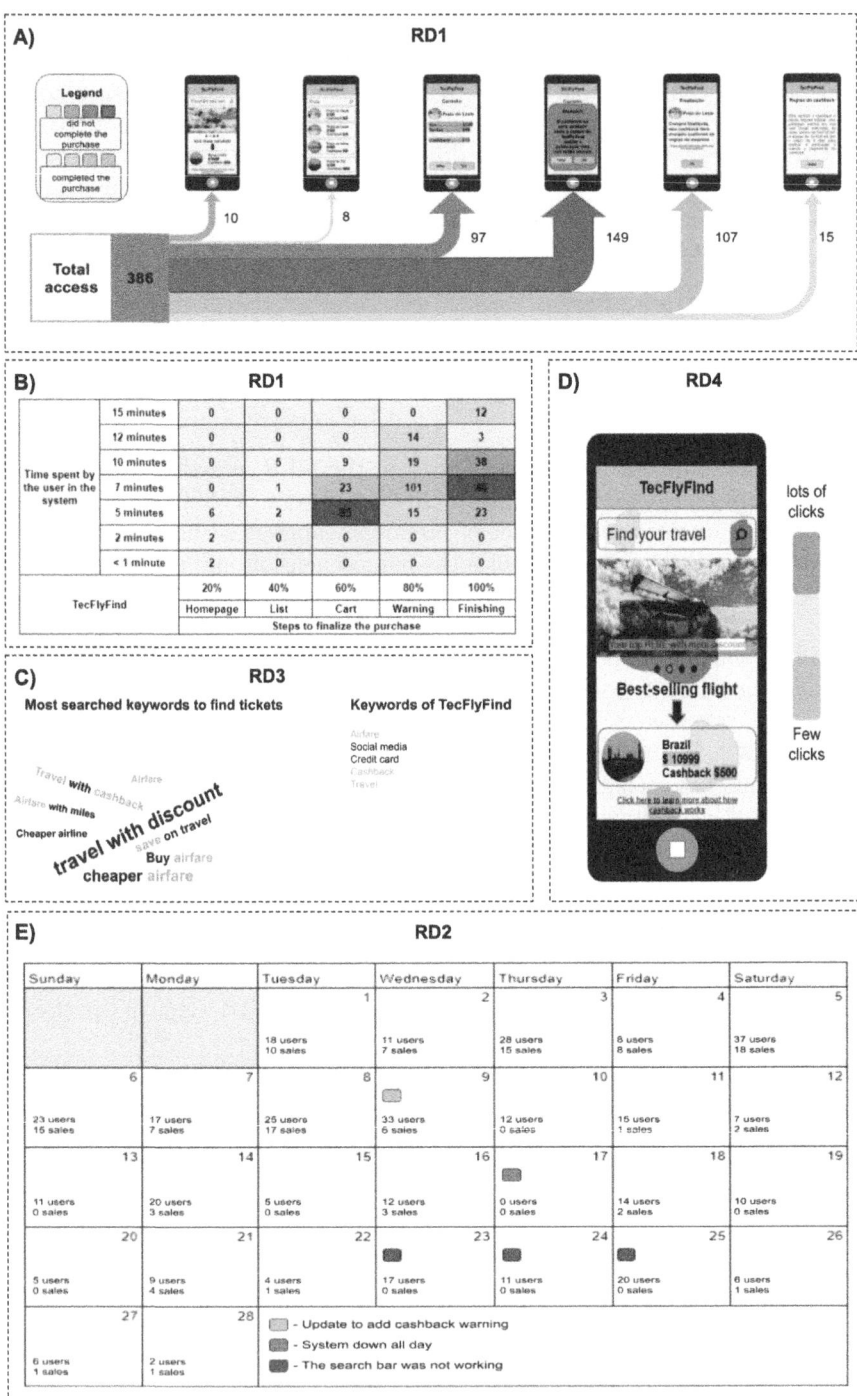

Fig. 1. Proposed visualizations to extract insights about reports of dissatisfaction.

– *How:* explore the different tools and techniques which can be used to analyze and visualize UX data, including data analysis tools, charts (e.g., diagrams and heatmaps), user journeys maps, user tests, interviews, and questionnaires.

The proposal's structure can also be a checklist to determine if the team has the necessary resources to implement a UX data-based approach. In scenarios where a UX data-based approach is already implemented, the framework's content can assist in identifying available resources and analysis opportunities. The final version of the framework is expected to serve as a tool for software teams to harness the potential of emerging UX data results within the development process, regardless of the team's background or maturity regarding UX data approaches.

Fig. 2. First framework draft.

New refinements will be applied in the first framework draft as the next step. The return to the GRL results as well as the exploration of the scientific literature and the primary study, will support the draft evolution. After evolving the draft, two evaluations are planned with at least two software teams.

For both evaluations, three moments of interaction are expected between the researcher and the team, including: (i) a training workshop on how to use the framework, (ii) the researcher participation in daily team meetings to observe the use of the framework, and (iii) assessment through interviews and/or questionnaires. The main objective of the first assessment is to analyze the teams' experience using the framework and determine what contributions and conditions are necessary for using the framework (e.g. context, requirements, resources). The focus of the second evaluation will be on acceptance of the framework.

8 Final Considerations

The progress of the project has already discovered: *Scientific literature:* definitions about UX data, the use of visualization to gain insights about UX issues, and research gaps about more flexible solutions for visualizing UX data. *GLR:* an overview of approaches that involve UX data (e.g., analysis, tools, and chart formats) and definitions about which professionals use UX data and what types of data are used. *1st Exploratory study:* the findings indicated the utility of the developed visualizations for designers, stakeholders, and developers. However, participants did not perceive that UX data could be collected without the involvement of UX professionals, and the definition of UX data remained unclear to software teams. *Framework:* the first version presents guided questions to be used as a checklist to determine the primary resources to implement a UX data-based approach and identify available resources and analysis opportunities. The evolution and evaluation of the framework are the future works of this project.

Acknowledgment. We thank Conselho Nacional de Desenvolvimento Científico e Tecnológico - Brazil (CNPq grant 309497/2022-1) for the partial financial support; and the Gary Marsden Travel Awards; and the participants of Doctoral Consortium of EICS 2023.

References

1. Adams, R.J., Smart, P., Huff, A.S.: Shades of grey: guidelines for working with the grey literature in systematic reviews for management and organizational studies. Int. J. Manag. Rev. **19**(4), 432–454 (2017). https://doi.org/10.1111/ijmr.12102
2. Albert, W., Tullis, T.: Measuring the User Experience: Collecting, Analyzing, and Presenting Usability Metrics. Elsevier Science, Interactive Technologies (2013)
3. Bakiu, E., Guzman, E.: Which feature is unusable? Detecting usability and user experience issues from user reviews. In: 2017 IEEE 25th International Requirements Engineering Conference Workshops (REW), pp. 182–187 (2017). https://doi.org/10.1109/REW.2017.76
4. Bernhaupt, R., Martinie, C., Palanque, P., Wallner, G.: A generic visualization approach supporting task-based evaluation of usability and user experience. In: Bernhaupt, R., Ardito, C., Sauer, S. (eds.) HCSE 2020. LNCS, vol. 12481, pp. 24–44. Springer, Cham (2020). https://doi.org/10.1007/978-3-030-64266-2_2
5. Buono, P., Caivano, D., Costabile, M.F., Desolda, G., Lanzilotti, R.: Towards the detection of UX smells: the support of visualizations. IEEE Access **8**, 6901–6914 (2020). https://doi.org/10.1109/ACCESS.2019.2961768
6. Card, S., Jacko, J.A.: Human-Computer Interaction Handbook: Fundamentals, Evolving Technologies, and Emerging Applications, 3rd edn. CRC Press Inc, Boca Raton (2012)
7. Card, S., Mackinlay, J., Shneiderman, B.: Readings in Information Visualization: Using Vision to Think. Academic Press (1999)
8. Desai, R.: 4 ways to implement data for a better user experience design (2019). https://www.freecodecamp.org/news/4-ways-to-implement-data-for-a-better-user-experience-design/
9. Dittrich, S., Hof, F., Wiethoff, A.: InteracDiff: visualizing and interacting with UX-data. In: Proceedings of Mensch Und Computer 2019, pp. 583–587. MuC 1919, Association for Computing Machinery, New York, NY, USA (2019). https://doi.org/10.1145/3340764.3344463

10. Fabijan, A., Olsson, H.H., Bosch, J.: The lack of sharing of customer data in large software organizations: challenges and implications. In: Sharp, H., Hall, T. (eds.) XP 2016. LNBIP, vol. 251, pp. 39–52. Springer, Cham (2016). https://doi.org/10.1007/978-3-319-33515-5_4
11. van Gaasbeek, J.R., Martin, J.N.: Getting to requirements: the w5h challenge. In: INCOSE International Symposium, vol. 11, no. 1, pp. 1083–1088 (2001). https://doi.org/10.1002/j.2334-5837.2001.tb02413.x
12. Garousi, V., Felderer, M., Mäntylä, M.V.: Guidelines for including grey literature and conducting multivocal literature reviews in software engineering. Inf. Softw. Technol. **106**, 101–121 (2019)
13. Hevner, A.: A three cycle view of design science research. Scand. J. Inf. Syst. **19**, 4 (2007)
14. Hokkanen, L., Kuusinen, K., Väänänen, K.: Minimum viable user experience: a framework for supporting product design in startups. In: Sharp, H., Hall, T. (eds.) XP 2016. LNBIP, vol. 251, pp. 66–78. Springer, Cham (2016). https://doi.org/10.1007/978-3-319-33515-5_6
15. Hokkanen, L., Väänänen-Vainio-Mattila, K.: UX work in startups: current practices and future needs. In: Lassenius, C., Dingsøyr, T., Paasivaara, M. (eds.) XP 2015. LNBIP, vol. 212, pp. 81–92. Springer, Cham (2015). https://doi.org/10.1007/978-3-319-18612-2_7
16. Homoliak, I., Toffalini, F., Guarnizo, J., Elovici, Y., Ochoa, M.: Insight into insiders and it: a survey of insider threat taxonomies, analysis, modeling, and countermeasures. ACM Comput. Surv. **52**(2), 1–40 (2019). https://doi.org/10.1145/3303771
17. Hussain, J., et al.: A multimodal deep log-based user experience (UX) platform for UX evaluation. Sensors **18**, 1622 (2018). https://doi.org/10.3390/s18051622
18. Karapanos, E., Martens, J.B., Hassenzahl, M.: Reconstructing experiences with iscale. Int. J. Hum Comput Stud. **70**, 849–865 (2012). https://doi.org/10.1016/j.ijhcs.2012.06.004
19. Kashfi, P., Feldt, R., Nilsson, A.: Integrating UX principles and practices into software development organizations: a case study of influencing events. J. Syst. Softw. **154**(C), 37–58 (2019). https://doi.org/10.1016/j.jss.2019.03.066
20. Kieffer., S., Rukonic, L., de Meerendré., V.K., Vanderdonckt., J.: Specification of a UX process reference model towards the strategic planning of UX activities. In: Proceedings of the 14th International Joint Conference on Computer Vision, Imaging and Computer Graphics Theory and Applications - Volume 2: HUCAPP, pp. 74–85. INSTICC, SciTePress (2019)
21. Lachner, F., Naegelein, P., Kowalski, R., Spann, M., Butz, A.: Quantified UX: towards a common organizational understanding of user experience. In: Proceedings of the 9th Nordic Conference on Human-Computer Interaction. NordiCHI 2016, Association for Computing Machinery, New York, NY, USA (2016). https://doi.org/10.1145/2971485.2971501
22. Melegati, J., Guerra, E., Wang, X.: Understanding hypotheses engineering in software startups through a gray literature review. Inf. Softw. Technol. **133**, 106465 (2021). https://doi.org/10.1016/j.infsof.2020.106465
23. Móro, R., Daráz, J., Bieliková, M.: Visualization of gaze tracking data for UX testing on the web. In: HT (Doctoral Consortium/Late-breaking Results/Workshops) (2014)
24. Munzner, T.: Visualization Analysis and Design. In: Peters, A.K. (eds.) Visualization Series (2014)
25. Norman, D., Nielsen, J.: Nielsen norman group - the definition of user experience (UX). https://www.nngroup.com/articles/definition-user-experience/ (2018)

26. Ohashi, K., et al.: Focusing requirements elicitation by using a UX measurement method. In: 2018 IEEE 26th International Requirements Engineering Conference (RE), pp. 347–357 (2018)
27. Petersen, K., Feldt, R., Mujtaba, S., Mattsson, M.: Systematic mapping studies in software engineering. In: Proceedings of the 12th International Conference on Evaluation and Assessment in Software Engineering, pp. 68–77. EASE 2008, BCS Learning & Development Ltd., Swindon, UK (2008)
28. Rivero, L., Conte, T.: A systematic mapping study on research contributions on UX evaluation technologies. In: Proceedings of the XVI Brazilian Symposium on Human Factors in Computing Systems. IHC 2017, Association for Computing Machinery, New York, NY, USA (2017)
29. Rodden, K.: Applying a sunburst visualization to summarize user navigation sequences. IEEE Comput. Graphics Appl. **34**(5), 36–40 (2014)
30. Shehabuddeen, N., Probert, D., Phaal, R., Platts, K.: Management representations and approaches: exploring issues surrounding frameworks. In: Bam, pp. 1–29 (2000)
31. da Silva Franco, R.Y., Santos do Amor Divino Lima, R., Monte Paixão, R.D., Resque dos Santos, C.G., Serique Meiguins, B.: Uxmood–a sentiment analysis and information visualization tool to support the evaluation of usability and user experience. Information **10**(12), 366 (2019)
32. Sultan, M., Miranskyy, A.: Ordering stakeholder viewpoint concerns for holistic enterprise architecture: the w6h framework. In: Proceedings of the 33rd Annual ACM Symposium on Applied Computing, pp. 78–85. SAC 2018, Association for Computing Machinery, New York, NY, USA (2018). https://doi.org/10.1145/3167132.3167137
33. Unterkalmsteiner, M., et al.: Software startups - a research agenda. E-Informatica Softw. Eng. J. **2016**, 89–124 (2016). https://doi.org/10.5277/e-Inf160105
34. Venable, J.: A framework for design science research activities. In: Proceedings of the 2006 Information Resource Management Association Conference, pp. 184–187 (2006)
35. Yang, L., Hu, Z., Long, J., Guo, T.: 5w1h-based conceptual modeling framework for domain ontology and its application on STPO. In: 2011 Seventh International Conference on Semantics, Knowledge and Grids, pp. 203–206 (2011). https://doi.org/10.1109/SKG.2011.31
36. Yang, S.Y., Chou, W.H., Chen, H.X.: Development of the automated tool to perform analysis and visualisation of user reviews. Int. J. Arts Technol. **12**(3), 218–237 (2020). https://doi.org/10.1504/IJART.2020.109814
37. Zaina, L.A., Fortes, R.P., Casadei, V., Nozaki, L.S., Paiva, D.M.B.: Preventing accessibility barriers: guidelines for using user interface design patterns in mobile applications. J. Syst. Softw. **186**, 111213 (2022). https://doi.org/10.1016/j.jss.2021.111213

Author Index

A
Angeloni, Maria Paula Corrêa 3
Artizzu, Valentino 151

B
Barathon, Benoit 41
Bisante, Alba 63

C
Calò, Tommaso 198
Campos, José Creissac 179
Carayon, Axel 28
Cau, Federico Maria 90
Chaabane, Sondes 41
Colin, Maxime Bourgois 41

D
Datla, Venkata Srikanth Varma 63
de Oliveira, Káthia Marçal 3
Dittmar, Anke 50
Dix, Alan 116

F
Fassi, Raja 41
Ferrai, Alice 41
Forbrig, Peter 20, 50
Freienstein, Jan 102

G
Gerken, Jens 102
Gómez-Monroy, Carla 166
Guerrier, Yohan 11, 41
Guffroy, Marine 11
Guilain, Dorothée 41

H
Hesenius, Marc 127

K
Kolski, Christophe 11, 41
Kronhardt, Kirill 102
Kühn, Mathias 50

L
Lebrun, Yoann 41
Lepreux, Sophie 41

M
Macedo, Maylon 207
Martinie, Célia 28, 72
Moreira, Ezequiel José Veloso Ferreira 179
Mourali, Yosra 41

N
Navarre, David 72

P
Palanque, Philippe 28, 72
Panizzi, Emanuele 63
Pascher, Max 102
Pudlo, Philippe 41

R
Ramírez-Reivich, Alejandro C. 166

S
Sauvé, Jason 41
Severin, Benedikt 127
Spano, Lucio Davide 90
Strugeon, Emmanuelle Grislin-Le 3

T
Trasciatti, Gabriella 63

U
Umlauft, Alexandru   20, 50

V
Vella, Frédéric   11
Vigouroux, Nadine   11

W
Werger, Ole   127

Z
Zaina, Luciana   207
Zeppieri, Stefano   63

SPRINGER NATURE

GPSR Compliance

The European Union's (EU) General Product Safety Regulation (GPSR) is a set of rules that requires consumer products to be safe and our obligations to ensure this.

If you have any concerns about our products, you can contact us on ProductSafety@springernature.com

In case Publisher is established outside the EU, the EU authorized representative is:

Springer Nature Customer Service Center GmbH
Europaplatz 3
69115 Heidelberg, Germany

The manufacturer's authorised representative in the EU is Springer Nature Customer Service Centre GmbH, Europaplatz 3, 69115 Heidelberg, Germany. If you have any concerns regarding our products, please contact ProductSafety@springernature.com

Printed and bound by CPI Group (UK) Ltd, Croydon, CR0 4YY
25/03/2026
02078187-0010